LIFE OF A CHALKSTREAM

SIMON COOPER

Dear Alastair,

Sending a touch of the chalkstreams north of the border!

Best wishes,

Simon

May 2014

WILLIAM
COLLINS

William Collins
An imprint of HarperCollins*Publishers*
77–85 Fulham Palace Road
London W6 8JB
WilliamCollinsBooks.com

First published in Great Britain by William Collins in 2014

15 17 19 20 18 16 14

1 3 5 7 9 11 10 8 6 4 2

A catalogue record for this book is available from the British Library.

ISBN 978-0-00-754786-9

Printed and bound in Great Britain by Clays Ltd, St Ives plc.

To Mary and Nigel. For endless encouragement and always being there.

For men may come and men may go,
But I go on for ever.

The Brook by Alfred Lord Tennyson

CONTENTS

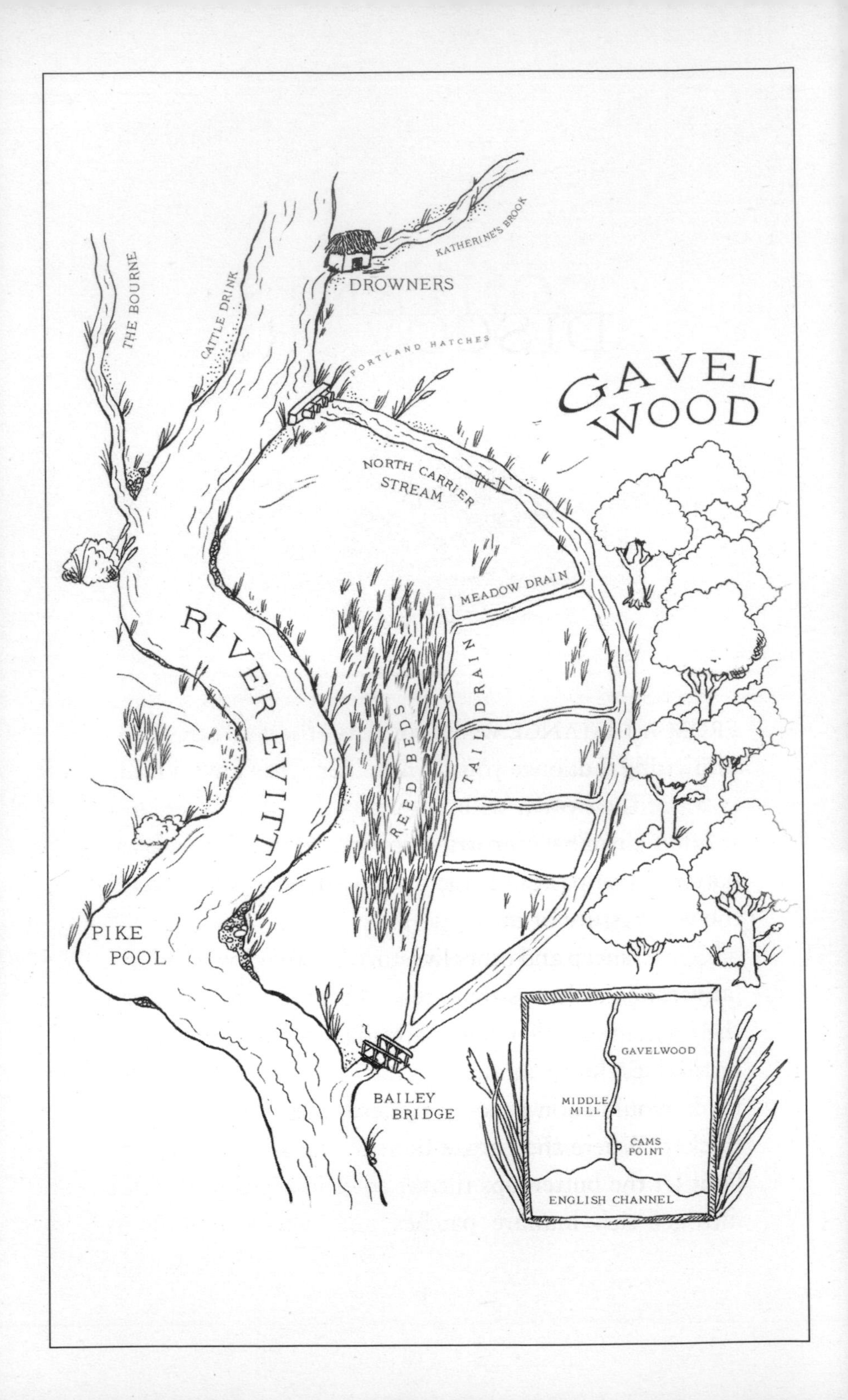

GAVEL WOOD
THE BOURNE
CATTLE DRINK
DROWNERS
KATHERINE'S BROOK
PORTLAND HATCHES
NORTH CARRIER STREAM
MEADOW DRAIN
TAIL DRAIN
REED BEDS
RIVER EVITT
PIKE POOL
BAILEY BRIDGE
GAVELWOOD
MIDDLE MILL
CAMS POINT
ENGLISH CHANNEL

1
DISCOVERY

FROM A DISTANCE water meadows look unkempt and uninviting, but once you get into them they have a beauty all of their own, with a myriad grasses, flowers and stunted shrubbery growing in an apparently irregular pattern. The pattern is dictated by the cattle that graze the wet pasture of the river valley.

Cattle, sheep and other livestock are the cloven-hoofed landscape gardeners that create the meadows. Without their relentless chewing, battering down the growth, fertilizing the ground and churning up the turf, the fields would soon become a dense, overgrown bramble thicket. Where they graze tight to the sod the sun and light let the buttercups thrive; cowslips spring from the nitrogen-rich manure patches and where their hoofs punch holes in the soil, the rhizomes of the yellow flag iris are split and separated to create fresh growth for the following season.

And sure enough, as I picked my way across the meadows I spied a diverse collection of cattle grazing in the distance, only their upper bodies visible above the pasture. Livestock are also great path-makers. Their sense of direction may be slightly off-kilter, and they may fail to realize that the shortest route between two points is a straight line, but they are canny and I know that if you deviate from the path they've trodden you will soon become stuck in boggy ground. So I followed their zigzag path across the field.

Reaching the cattle, a motley collection of brown and white Hereford crosses, black Aberdeen Angus and the pale, long-limbed, lean continental types, I paused to consult the map. The cattle paid me little interest, raising their heads now and then to check me out, but never pausing as they masticated their way through their daily mass of roughage. I'm told that meadow-grazed beef is the sweetest, most tender meat of all but it seemed unfair to share this news with them.

To my right the summer brown of the grassland gave way to a vivid green ribbon, the best indication yet that the river was close by. The dry cattle path petered out, giving way to wet ground poached by a thousand hoofs where the cattle had grazed right up to, and under, a barbed-wire fence. In fact the grass immediately under and just the other side of the fence had been grazed as tightly as a bowling green; proof that – for cattle at least – the grass is greener on the other side of the fence.

Picking the stoutest fence post, I climbed onto the top strand and from this vantage point caught my first view of a sparkling river. I was still separated from the river itself by 30 yards of rushes and as I leapt to the ground the other side I sent up two silent prayers of thanks. First, that I had had the foresight to put on waders – those 30 yards were likely to be slimy, smelly and difficult to negotiate. Second, that the river was fenced, because cattle and rivers simply do not mix. Give cattle a chance to graze right up to the edge of the river and that is what they will do. However, cattle are big, clumsy beasts that don't mind getting their feet wet in search of that extra special, tasty mouthful. So they yomp up, down and along the edge of the river, gradually destroying the banks and vegetation.

Imagine you have a fenced river corridor that is 50 yards wide. In the middle you have 20 yards of river, a width that the river has arrived at more or less of its own accord to accommodate the variable winter and summer flows. Either side of the river lie 10 yards of semi-aquatic vegetation: plants like rushes, watercress and wild mints that like to live half in and half out of the water. This wet area is the perfect home for the insect life that will ultimately sustain a fly-fishing river. The final outer five yards on either side will be hard bank that contains the river in all but the heaviest of flood conditions and is home for the sedge grasses and tussocks that like their feet dry for most of the year. So far so good. Now take away the fencing. Within a matter of hours the cattle will discover this new Elysium and within a few days the wetland greenery will have been grazed to water level. Not content to leave it alone, the cattle will persistently graze the new shoots. Their strong legs and sharp hoofs will destroy the root structure, slowly killing the plants from below. The first winter flood will wash away the soil, exposing the gravel bed below, and the plants will be unable to re-establish themselves in the faster water. Within a short time the river that was once 20 yards wide is now 40 yards wide, shortly to become 50 as the cattle destroy the hard bank as they lumber in and out of the water. Having a river that's two and a half times wider might not seem such a bad thing, but assuming the volume of water stays the same, which it will on a chalkstream, the depth will be two and a half times less, and for a trout at least, this is bad news on every level – food, survival and breeding. If you are a trout hanging out in your favourite spot close to the bottom, looking upstream into the column of water above you for stuff to swallow, then the greater the depth, the greater the choice of food, which is why trout tend to gravitate to the deepest pools unless they are in search of particular food or get chased out by bigger trout.

Always on the lookout for food, trout are wary creatures that have plenty of predators. When they are small the greatest danger is other trout or maybe kingfishers, but as they grow larger pike, cormorants, herons and ospreys, otters and mink are ever-present dangers. In every case, except for the smallest of fry, the deeper the water the less likely these threats are to attack the trout, and if they are attacked the depth gives more options for escape. Trout fry on the other hand like to hide out in the reed beds either side of the main channel. More practically, for the survival of the species trout need to lay their eggs in loose gravel that is constantly washed with rapidly flowing, well-oxygenated water that percolates down to the eggs. Take away that speed of flow by spreading it across two and a half times the width and suddenly too little good water will flow over the eggs and they will slowly die due to lack of oxygen.

However, standing just past the fence contemplating striding out across 30 yards of swampy reeds to reach the river I was more concerned for my safety than with any ecological niceties. I have learnt from bitter experience that the worst thing to do is to adopt a bold Neil Armstrong-like moon stride – all that will happen is that your leading leg will disappear into the mud, upending your face into the slime. Far better to shuffle forward, letting the weight of your feet break through the surface and allowing you to sink slowly until you reach firm bottom;* then it is a question of somehow

* I have no hard or fast rule about what to do if you don't reach hard bottom. Generally if I am still sinking when the gloop reaches my waist I rapidly turn tail to heave myself back onto the firm bank with a fair imitation of an arthritic walrus.

walking/shuffling/pushing your way through the mire with reed roots grabbing at your feet. Each movement that disturbs the mud releases a noxious smell: part methane, part rotting vegetation, part musty odour. Sometimes your passage will bring an oily slick to the surface. And unpleasant though that might be, it does demonstrate what a huge natural filter the river's edges provide, the excessive nutrients and run-off degrading in the mud rather than being washed directly into the river.

Hindsight suggests that I had not picked the easiest place to get into the river. As it turned out, a few hundred yards upstream the reed margin narrowed to a few feet, but as I stepped out of the reeds I was in the most perfect river, and at that moment it was worth the effort. A fast, clear stream with huge rafts of waving green crowfoot, which is essentially water buttercup with a white rather than yellow flower, filled the river, the gaps interspersed with bright gravel patches. Donning polarized sunglass-es to cut out the surface glare I began to scout the depths of the water, picking out the occasional brown trout in the open water and disturbing shoals of grayling as I waded upstream.

Seeing the trout made me happy, but seeing the gray-ling happier still – not so much from an angling view-point but because grayling are an indicator species that confirm the good health of a river. They are far more sensitive than trout to declining water quality, and if they disappear you know you are in for problems. They are not so much the canary in the cage that drops dead when the danger has arrived, but rather the bird that flies away at the first sniff of trouble. As for salmon, my suspicion was that I would see them in the autumn; the Evitt has a reputation for a run – an influx of fish from the sea – that comes in late autumn to spawn, but for now that was simply conjecture.

As I pushed on up the river the morning began to warm up and after a while a hatch of olives appeared above the water. 'Olives' is one of those words fishermen bandy about. It is a catch-all name that describes a whole range of insects that are to be found flying on the river, going about their daily business of survival and procreation. They are important to anglers because olives are one of the staple foods in the trout's diet.

I say 'appeared' because it always seems to be that way – one minute there are no insects, the next there is a cloud gathered above the water or alongside the water. For the chalkstream fisherman the sight of a hatch is a promise of things to come, because eventually when those insects alight on the surface of the river, either to lay their eggs or to die, hungry trout will eye them up, rise to the surface and swallow them down along with a gulp of water.

The very essence of dry fly-fishing, dating all the way back to the Macedonians around the time of Christ, is to imitate this process. Take a hook, decorate it with fur and feather to create a fake that looks like the real fly. Tie the hook to the end of your line and then use a rod or cane to cast the fly onto the water so that it lands like thistledown on the surface, thereby imitating the natural landing and fooling the trout into mistaking it for food and making a lunge for it. If all goes according to plan you raise the tip of the rod, tighten the line and set the hook into one very surprised, and soon to be furious and fighting, trout.

In fishing jargon, this is referred to as 'matching the hatch' – observing the insects on which the trout are feeding and fishing the artificial imitation. Spend time in the company of anglers reporting back from a day on the river or read the comments in the catch record book and you'll get a sense of how knowledge of entomology, rudimentary, encyclopedic, or just plain guesswork, dictates the pace of a fishing day. You'll come across phrases like 'a great hatch of olives', 'plenty of blue-wings about' or more honestly, 'couldn't really make them out – maybe some sort of small olives?' You will nod your head wisely but will most likely be none the wiser at all and put it down to some riverine double-speak. In an idle moment you might even pause to wonder what this much spoken about 'olive' is, but move on quickly – you probably have a life to live.

Actually the truth is you have probably seen olives on thousands of occasions without even registering their existence, for *Baetis*, to give them one of their more common Latin names, inhabit just about every lake, pond and river in the British Isles. Next time, look out for a small cloud of insects, hovering just above or beside the water – they are certainly some kind of olive that hatch through spring, summer and autumn. An individual olive will look like a round bundle of fur fluttering on the air, keeping in time and close proximity to the hundreds of others, all identical. In fact olives are not round at all, they just look that way, as their wings are a blur to the human eye, beating thousands of times a minute to keep them aloft.

If you can ever get one to alight on your hand they are creatures of the most extraordinary beauty: big black eyes, impressive mandible, large translucent, veined wings and long triple tails shaped like a cat's whisker that double the length of their tapering, segmented body. In angling parlance they exist as large, medium and small. Large is the size of a blueberry, medium a pea and small an unsplit lentil. As the name suggests, they are olive-coloured or a drab green of varying hues. Sometimes the wings differ in colour from the body, which gives rise to types such as the blue-winged olive, but anglers like to keep their nomenclature simple and to the point, if a little dull. But that said, the blue-winged olive has a hint of the exotic about it, and the claret dun a gravity that suggests it must succeed.

'Dun' – there's another word that creeps out of the angling lexicon, but what on earth does it mean?

Essentially the insects you see in the cloud by the water are at one of the latter four stages of life – egg, nymph, dun and spinner. The first two stages take place in the water, mostly out of sight, while the third and fourth are played out in the air for all to see.

The dun is the olive you can see hovering above the water, flapping his or her wings for all he is worth as he keeps up with the pack. He is, in human terms, a maturing adolescent, just a few hours or at most days old. The pack instinct is part mating ritual, part holding pattern while the body matures and morphs into the next stage: the spinner. Even in the world of drab olives, becoming a spinner equates to a new level of attractiveness – your tails get longer (truly!) and you'll be a much brighter colour than your previous dun camouflage. This heralds a brief flurry of sexual activity.

Spinners. I have no idea how they got their name. Maybe it describes the mating dance when the pair flies up in unison and then hovers for a moment at the top of the climb before relaxing their wings to spin down on the air. Maybe it is because in olden times people thought the long trail of eggs was something akin to spinning yarn. Or maybe it is the dead insect circling on the current. Whatever the reason the female spinner, ready to lay her eggs, is brighter than in her maiden form. Clearly the consummation brings colour to her wings and body. The egg-laying is a bittersweet moment to watch. On the one hand it is the proof that a new generation is on the way, but on the other that the insect will be dead in a matter of minutes or a few hours at most.

Some days on the river I will see one type of insect to the exclusion of all others, but today was one of those days when the diverse population was out in force; good news for the ever-hungry trout. There is not a lot of nutrition in a tiny insect, even for a trout, so it's all about the effort/reward equation. A huge fat mayfly – the size of a dandelion head – is worth that extra effort, but the tiny corpse of an olive a gentle slurp. Somewhere in between is the impregnated female, stuffed with energy-rich eggs. The latter is so attractive to fish that fly tiers will add a tiny wrap of yellow thread to the underside of a fly – no more than an eighth of an inch long – to represent the egg sac.

There are dozens of species of fly to be seen. They make their lives on the river, but ultimately the eggs will be laid in one of two ways: on the surface or beneath it. For the angler and casual observer it is the surface layers that are the most interesting, especially the sedges. Sedges, or caddis, are big flies in the general run of a chalkstream. Not as big as the mayfly, but four or five times the size of your average olive. They are very much summer creatures, present beneath the current all year but hatching only in June, July and August. If they look like anything else, it is the common household moth with its wings folded in to create a tent over the body. If that sounds clumsy you would be right. Sedges are clumsy; the worst fliers and worse still at landing. Their approach to the river surface will look fine, but come the final few inches, instead of swooping gently down to clip the water to allow the surface tension to draw the eggs from her body, the female caddis will crash onto the water. Alerted by the commotion, trout from many feet away, even facing in the opposite direction, will turn and make a grab for the egg-laden wreckage. The smaller olives are, by comparison, incredibly delicate, getting within a fraction of an inch of the water before depositing their eggs.

For the angler tying on a sedge imitation this is a moment sent from heaven. There's no delicate cast required here. No, a splashy cast will do as well, if not better, and the eager trout will do all the work to grab the fly. The olives are a different matter. You will need your thinnest line, your tiniest fly, your most accurate and delicate presentation. And even when you get it perfect, the languid trout, with time to weigh up all the options, will as often as not reject your offering.

It is relatively easy for the sub-surface egg-layers to go about their business unobserved, but the big problem is getting through the surface tension of the water. An insect with wings is quite bulky; it has a large surface area that is gripped by the water. Just sitting on the top and hoping to paddle their way underwater will not work. They need purchase and they find this from the reeds, stones and tree roots emerging from the water. As I pushed upstream on that July morning it was the perfect time of year for the blue-winged olive. And sure enough there they were with their drab olive bodies and translucent blue wings, arrayed along the length of the upright dark green reeds that gently swayed in the margin. Unfortunately there is an unusual predator that lies in wait.

As I watched the olive closest to the water edge down towards the film, and as she forced her body into the water, I could see the six tiny legs straining on the reed, the little suction caps on the feet giving her the leverage required. But in this moment of supreme effort it is the misfortune of the olive that the European eel chooses this very time of year to begin its downward migration to the sea. After ten or fifteen years in a muddy pond *Anguilla anguilla* heads for the Sargasso Sea, but before the ocean the river provides a welcome source of food. In the shade of the reed it is hard to see the eel going about his business, but in the early morning or late evening you will surely hear them. It is a slurping sound, a bit like a child sucking up the last of a milkshake with a straw, as the eel quite literally sucks the insect into his mouth at the very moment it is caught by the surface tension.

Fortunately there are many more olives than eels to consume them, so very soon the sunken spinners are laying their eggs beneath the surface. These then drift slowly down on the current to lodge in the stones, silt and general debris of the riverbed where they will remain for week or months until they become nymphs and embark on the next stage of life. I am not sure if the spinners themselves are able to hold their breath or even breathe underwater, but it probably does not matter. The time is short between submersion and being spent, namely eggs laid and becoming a semi-lifeless body, tumbling downstream on the current. The spinners that lay on the surface fare no better, collapsing exhausted on the surface, the job done. At first they lie on their sides, with one wing up, but as the life seeps away the other wing collapses and the end finally comes with convulsions that cause the water to ripple outwards around the insect until it stills.

Trout are no respecters of death, and sure enough, just off the main current, in a back eddy I came across a confident trout cruising in the slack water. With his back out of the water and his body submerged to eye level he languidly circled around, his mouth open, the flow of water carrying the spent spinners down his throat. This is the ultimate effort/reward equation and he keeps at it until the surface is cleared. Above him the duns, newly hatched, buzz in the air but he pays them no attention and the insect mortuary empty, he fins down to the deep to await the next funeral cortege.

A chalkstream in summer – June and July – is when it is most alive. It seemed that every step I took that morning, in the river or on the meadows, brought a new discovery. Above the shallows, on a dead branch, a kingfisher waited impatiently for the fry to move into the shallow water as it was gradually warmed by the morning sun. I am not sure kingfishers are really impatient, but the way they cock their head back and forth makes it look that way. I am certain the head-cocking is just to change their angle of vision so that they can see through the surface glare to catch sight of the fish, but for whatever reason, once locked in on the fry a rapid blue streak flashes from branch to water and back again in an instant. Holding the fish crossways in his beak the kingfisher raises his head, straightens his neck, turns the fry head-first and swallows it whole.

With a shake of his feathers, the watch will resume. This is likely to be an all-day affair, because the more the sun shines, the warmer the shallows will become and the more fry will appear in darting shoals. And the kingfisher is on a mission to feed. Somewhere along the bank, in a spot I was yet to locate, was a nest burrowed into the soft soil. In that nest would be maybe up to half a dozen chicks, each of which needs a dozen or more fish a day. That is getting on for a hundred fish. I watched our impatient friend catch four more and then left him to it, making a wide circle around the shallows to leave his hunting ground undisturbed.

By now the geography of the river and the meadows was starting to make some sense, and as I waded upstream the structure of the place began to arrange itself before me. The main river was the spine. Coming in from the left was a bourne, a small stream that only flowed in any significant sense during the winter and early spring, so by now was near to dry. It would remain so until the autumn rains. Cutting off at a sharp angle to the right, heading due north for most of its run, was a carrier, a channel dug by hand many centuries ago whose sole purpose was to flood the meadows from February to May. Now abandoned and choked with overgrowth, the carrier was clearly once pivotal to the water-meadow system. As the channel that moved the water out of the main river across the meadows it had leats or ditches that ran off at regular intervals on both sides as conduits for carrying off the water to flood the fields. But in an arid July the leats were dry and hidden by the summer meadow grasses. Come the winter they would reveal themselves.

As I pushed on up the river the sun burnt off the cloud; it was getting warm but I still had two more things to find: the Drowners House and the brook. Wading in gin-clear water under an azure blue sky is hardly the toughest job in the world, especially in a chalkstream like the Evitt that has no great incline to it. It doesn't race like a tidal river or rush in torrents like a mountain stream; rather it glided across the face of my waders at a gentle walking pace. Looking upstream from where I was standing I could see a full 300 yards of river ahead, and I'd be hard pushed to swear that I could see a difference in height. In fact I knew that the source, 30 miles from the sea, is only 85 feet above sea level, so that is no more than two inches' drop in every hundred yards. For potomologists – those who study rivers – this is just about as benign a flow as a river can have.

This is a floodplain that is almost as flat as the river that flows through it, and it was only in the far distance, at least 3 or 4 miles way, that I could see the sheep-grazed downs that gradually rose to a few hundred feet. Long, long ago, in the ice age, the river valley was carved as a shallow ravine, but gradually, over millennia, the water flowing to the sea had left soil, silt, gravel and sand behind after the floods of winter, creating the flat plain on which the water meadows sit today. But nature did not do this all alone; man played his part. It is the conjunction of water with meadows that makes this such a very special landscape. There are meadows the world over, but in very few places has man harnessed the seasonal floods to irrigate and protect the grassland for the sole purpose of making the sward grow faster, lusher and more nutritious for cattle to graze. This ancient agricultural practice has, by chance and unintended consequence, created a home for a unique collection of creatures that are my constant companions.

Through a gap in the reeds I thought I spied what looked like the Drowners House some way across the meadows, with a bedraggled thatched roof covered as much by wild grass and weeds as by darkened straw. Heading for the gap to haul myself out of the river I crossed the path of a water vole swimming fast along the edge of the reeds, hugging the margin for protection. For such small creatures, seemingly so ill-adapted to water, they really can swim fast. In the water there was no way I could keep up in my waders, and on the bank I'd need to maintain a brisk walk as they stretch out their brown furry bodies, nose poked up in the air while their legs paddle like fury.

But this one, in common with all water voles, can only keep up the furious burst of speed for a short while. Quite suddenly he stopped, gave me a look with those little black eyes and with a plop disappeared beneath the surface. Under the surface things are a mad scramble for the water vole. With all that waterproof fur they are naturally buoyant, and their tiny lungs are not suited to holding their breath for long. But he had chosen to dive at this spot for a purpose. Beneath the water he wove between the roots of the reeds, heading for the bank. I could track his progress by the muddy trail he was leaving in the water until he reached the entrance to the burrow. At the tiny hole – no bigger than the size of an egg, just above water level and shiny from constant use, he stretched out the hand-like claws of his front legs, pushed them into the soft soil and using the purchase, squeezed himself inside the burrow.

Like the kingfisher, these are the breeding months for our water vole (*Arvicola amphibius*), which is now probably into its third or possibly fourth litter of the year, having started back in March. In the burrow, lined with dried grass torn and gathered from the bank above, anywhere between five and eight tiny voles, no bigger than your thumb, will be mewling for food. Back and forward go the adults, for anything up to eighteen hours a day. Fortunately they are pretty promiscuous in their diet on the herbivore scale. Little tooth marks on the reeds and sedges are easy to spot. Wild mint and watercress are chewed with relish, but of all the things it is wild strawberries that they fall on like mammals possessed. But really they will eat anything; the family demands it.

Doing the maths, you'd think that we'd be overrun by water voles by June – after all they are not great travellers and this one nest will have produced fifteen offspring by now. The truth is that being born a water vole is a high-risk incarnation. First, the weather might get you: lengthy bouts of bad weather, or worse still an ill-advised burrow that gets flooded. Inside the burrow you might be generally safe, but the common brown rat or worse still a stoat or mink will make short work of you and your family if you're discovered. Once outside you are assailed from above and below: owls, buzzards, otters and pike are just four of the predators who see you as a tasty morsel. If you make it to the semi-hibernation of winter you have done well.

By now it was getting a bit hot for trudging across rough meadows in waders, so I took a direct line to the Drowners House, stumbling on the way into what were most likely carrier ditches, still soggy at the bottom beneath the tangled grasses. Ducking down under the oak lintel of the doorless entrance I entered the cool of the house. With no windows it took a few moments for my eyes to adjust to the dark, while some streaks of light came through the holes in the dilapidated thatch, illuminating the river that ran beneath the ragged floorboards. A house with a river running through it for no apparent purpose? It could only be for the drowners.

The drowners are long gone, the purpose for their livelihood disappearing when modern agricultural methods consigned the water meadows to history. But for four centuries these were the men who regulated the flow of water from the river, through the drains and carriers dug across the meadows, to quite literally 'drown' the late winter and spring grasses in water. Warming the soil and air of the meadows – grass grows at 5°C, chalkstream water is 10°C – plus all the nutrients the water carried with it, was the perfect way to get cattle grazing earlier and thus create heavier crops of hay. Of course all this came at a price in terms of working conditions. Obviously the times when the water levels needed most adjustment, hourly and daily, came when the weather was most foul, so the drowners built these houses. They built them over the water for the same reason they drowned the grass – warmth.

I didn't need to take out a thermometer to check the temperature inside the house today; I knew it would be exactly the same as the water – 10°C. The thick triple-skinned red-brick walls, damp from the foundations in the wet land, helped keep the place the same temperature all year round. Today I was grateful to find it a full 10° cooler than in the sun outside, but I doubt nearly as grateful as the drowners were when the mercury fell below freezing in winter. Today, other than the house itself, there is not much evidence of the drowners' tenure. There are soot-blackened nooks hollowed out in the brickwork as candle-holders and some initials carved in the oak beams, but the current residents are mostly house martins that have coated the walls with white guano from their nests in the rafters above.

I found myself a handy log, placed it up against the hut wall and sat down to contemplate my options.

To say the river and meadows were in crisis would pitch it too strong. Severe neglect was closer to the truth; a river caught in a spiral of decline. For all the beauty of the river and the wildness of the meadows the creatures were in retreat. With every year that passed the spawning grounds were growing fewer as the streams and carriers progressively became blocked. Along the banks the scrubland was encroaching, eliminating the wide open spaces that natives like the water voles require. In the meadows, without proper grazing, the meadow plants were being crowded out. Untended, the clear, fast chalkstream waters of the floodplain would revert to a swampy morass, with insects like the olives and mayflies disappearing.

It could be saved, but was it worth saving? The answer had to be yes. The question now was how.

2
DECLINE

THIS MORNING I found a bat caught on a hook that was dangling from a snagged fishing line on a branch overhanging the river. This is not the first time I have found bats snared like this. Bats with their super sonar hearing home in on a discarded fishing fly mistaking it for a real insect and wham, they are impaled on the hook. Sometimes by the time I find them they have died, but this Daubenton bat, the species that most commonly populate the river valley, was definitely alive and very angry.

I have heard it said that the Daubenton is the only British species to carry the rabies virus. I have no idea whether this is true, but I don't intend to be the one to find out, so taking my handkerchief I swaddled the bat before snipping the line. Angry does not adequately describe how the brown, furry bat looked at me. The tiny black raisin-like eyes glared at me in pure fury. The pointed leathery ears that indicate the mood of the bat from gently lying back on the head (content) to being at rigid right angles to the head (agitated) were most definitely the latter. As I walked back to the Land Rover to get a pair of forceps to extract the hook from his belly I could feel his bony body twitch and turn in my hand, his head swivelling in an effort to locate the best direction of escape.

Bats have a bad press, but it is hard to feel anything but sympathy for the Daubenton for a moment or two. Though he seems exceedingly ungrateful for my help, at bay the furry head is more reminiscent of a mouse and the pink face, with wisps of downy hair, baby-like. That said, when he opens his mouth to snarl he exposes a vicious jaw full of sharp, ridged incisors designed to crush prey with one bite in flight. At the Land Rover, using an additional cloth I cover his head, trap his wings and expose the underbelly. On his back his hind legs, with claws like a bird but razor-sharp and bristly, struck wildly at the air, trying to get some purchase. The hook was caught in the belly, plumb between the legs, which made some sort of sense, for bats grab for their prey in the air with their feet. Grasping the eye of the hook with the nose of the forceps, I deftly twist my hand to remove it with one smooth movement. I am sure the Daubenton had absolutely no idea what was going on as I shook out the cloth to allow him to fly away. But he seemed to be none the worse for the experience and headed for his roost in one of the trees close by the river.

The Daubenton bats get to become regular companions if I hang around the river late into the evening anytime from May to September. The first few times I see them in May I get to do something of a double-take as the small, unfamiliar black shapes zip around the air. By now in September they are part of the furniture, and I can set my watch by them, as they appear almost exactly ninety minutes after dusk each evening. They are voracious feeders of chalkstream insects; midges are a particular favourite as they can swoop through the clouds of chironomids that gather above the river surface on calm evenings. Sometimes the bats will even take the hatching midge pupa from the surface, trawling their hind legs through the film. I am guessing it is those bristles on their feet that 'sweep' up the insects from the water that allow them to do this.

The bats patrol the air close to the river, high above the trees and everywhere in between for hours on end each evening for food, not just singly but in groups appearing from the trees closest to the river where they roost during the day. They sometimes, but not often, make a little squeak in flight. It is often described as a click but it never seems that way to me, but rather like the modulated squeak from a dog toy. But soon I will hardly see or hear them at all as they mate, become solitary and spend the winter in a safe roost.

September, the month of autumn fruitfulness, is a time of departures and preparations – everyone and everything in the river has its way of taking nature's cue of the impending winter. The adult swans thrash up and down the river to chase away their cygnets, creating chaos for anglers and other birds alike. After a few days the cygnets get the hint and take flight. Woe betide any youngster who tries to return. The male cob swan will have no qualms about a full-on attack, mounting the much smaller cygnet, biting his neck, smashing down with his wings and pushing the young bird beneath the surface until the point is made. The departure of the swallows is altogether a more orderly affair, daily gathering in greater and greater numbers until quite suddenly one day they have gone on the long migration to South Africa, to return in April. Along the banks the water voles revel in the autumn harvest of hazelnuts, blackberries, seeds, acorns and whatever else falls to the ground. In the meadows the farmers put out the cattle to get the last and best of the grazing. In the river the trout, sensing the onset of autumn by the shortening days, start to feed in earnest on a spectacular array of insects that hatch in great numbers to capture the last truly warm days of the year.

For fishermen September is often termed the month 'the locals go fishing', on the grounds that it is the best month and best-kept secret in the piscatorial calendar. But maybe, like the creatures, we anglers also sense another season drawing to a close and get just a little frantic to enjoy the last of it before the bar comes down. For me it is always the first flurry of autumn leaves blowing onto the surface of the river that tells me the end of the season is around the corner. If I am fishing it can be a little annoying, difficult to pick out my fly amongst the blow-ins, but whether I'm fishing or just walking the banks, the sight of dead brown leaves makes me sad for the end. But this year I am buoyed by the plans we have to restore Gavelwood, which will start immediately the fishing season has closed and continue through the winter.

Gavelwood, the land, the river, side streams, brook and water meadows, takes its name from a wood that makes up part of this tiny, forgotten part of England. The woodland, a mixture of native trees like oak and ash, is as unkempt as the meadows it borders. Nobody knows where the name came from, but it is clearly marked on the deeds of ownership. The medieval word 'gavel' meant to give up something in lieu of rent, so maybe in some distant century the lumber was exchanged for tenure. But here today I am not here for any timber, it is the river that is the draw. A beautiful chalkstream called the Evitt that runs gin-clear, the perfect home for fish and water creatures that thrive in a habitat that is as endangered and as worthy of protection as any tropical rainforest or virgin Arctic tundra.

The water that flows through the chalkstreams is a geological freak of nature, almost unique to England. There are a few chalkstreams in Normandy, northern

France, and one is rumoured to exist in New Zealand, but taken as a whole 95 per cent of the planet's supply of pure chalkstream water exists only in southern England. The water I watch flow by in the river today fell as rain a hundred miles to the north six months ago, was deep underground yesterday and will be in the English Channel in a few hours' time, a cycle that has been repeating for tens of thousands of years since the last ice age ended.

A chalkstream river valley today is a tamed version of how it started out. After the ice age it would have been little more than a vast, boggy marshland, with no river to speak of but rather thousands of streams, rivulets and watercourses that randomly flowed this way and that. At some point in time, it is hard to say exactly when, the early Britons must have started to use the valleys for a purpose, initially farming, which involved draining the land. Inevitably drainage involved reducing the myriad streams to a few channels, which in turn became the rivers that have evolved into the chalkstreams we have today.

It has been a mighty long process: five or six millennia for sure. The barges that carried the stones for Stonehenge were brought up what is now the Hampshire Avon, probably widened and straightened for the purpose, from where it enters the sea at Christchurch Harbour 33 miles from Amesbury, the Avon's closest point to Stonehenge. But these incremental activities changed the river valleys very slowly, and it was the advent of the watermills that was to prove the penultimate step on the way to the chalkstream valleys we see now.

Again it is hard to pinpoint precisely when watermills became a regular part of the landscape. One thing is for sure, there are plenty listed in the Domesday Book, so it is fair to assume that the valleys were taking shape to meet the requirements of water power by this time. Essentially the mill wheel requires a good head of water to drive it, so a special channel would be dug to supply the water to drive the wheel. This 'millpond' would be controlled by a series of hatches, which when opened would turn the wheel for a few hours. Once depleted, the hatches would be closed and the millpond given time to refill from the river and streams.

The unintended outcome of all this would be to drain the land in the immediate vicinity, which in turn created the most wonderfully rich grazing pasture on the alluvial soil left behind after many millennia of flooding. This bounty of nature did not go unnoticed, so over the centuries that followed the river valley was gradually drained not just for the mills but for farming. The water was concentrated into a single channel which is the River Evitt today, supplemented by the side streams and ditches that provide the drainage.

But the story has one last twist. Having deprived the land of the flooding, the farmers realized that they were taking away one of the very things that had made it so productive in the first place – the nutrient-rich water that every winter washed over it. So around the seventeenth century, as the agricultural revolution took hold, landowners realized that drainage alone was not the answer and that managed flooding would dramatically increase the yield from the land, so the water meadows came into being.

By digging carriers, or leats, quite literally streams that carry water away from the main river, redirecting side streams, filling in others and creating a series of hatches to manage the flow of water, the farmers were able to use the winter and spring flows to flood the meadows from February to May. The term flooding is something of a misnomer; deep, static water over the grass would do little more than rot it away. The skill in floating, the creation of a water-meadow system, is to keep a thin layer of water constantly moving over the surface. The warmth of the water and the protection from frost, plus the nutrients carried in from the river, allow the grass to grow earlier and quicker. When ready for grazing the cattle would be let in, to be taken off when they had eaten it down and the land reflooded. If this all sounds a laborious process, it probably was. It was far beyond the daily regime of the farmers who banded together to employ a drowner, or waterman, who regulated the flows.

Today drowners are a long-distant memory, the advent of artificial fertilizers sounding the death-knell for the meadows from the early 1900s. When the watermills finally stopped grinding a few decades later, the *raison d'être* for this integrated water system would have all but disappeared except for the fact that somewhere along the line, in the period when the chalkstream valleys went from marshes to meadows, the brown trout had become the dominant species in the river. Never ones to miss an opportunity, anglers soon followed, first for food and then for sport, at which point the chalkstreams became a byword for angling perfection. The drowners and farmers were replaced by river keepers who lavished care on the rivers far beyond the basic needs of an agrarian England.

Fishing, angling, call it what you will, with an insect, worm, net, hook, spear or anything else that captures the fish, is as old as mankind. But as a pastime, done for the pleasure of the activity as much as for the outcome, it has to be credited to the Victorians. They did of course have their antecedents. Dame Juliana Berners, an English nun, wrote *A Treatyse of Fysshynge wyth an Angle* in 1496, which can be claimed as the first book about fishing as a sport, although she has been eclipsed in history by Izaak Walton's *The Compleat Angler*, which followed 150 years later. But these great anglers and writers were exceptions; for most people trout were there for catching and eating with the minimum of effort. So why the Victorians? Well, it was a coming together of wealth, leisure time, technology, the railways and the insatiable curiosity of a few individuals.

Gavelwood today is a tiny proportion of what was once a huge country estate, running to thousands of acres and 11 miles of the River Evitt. In fact the entire river valley, encompassing all 30 miles of the Evitt from source to estuary, was in the ownership of just three families. Hardly very egalitarian, but those were the times, and for fishing, and the chalkstreams in particular, they proved decisive for the future. Once the fishing craze caught on amongst the landed gentry the rivers became much more than farmland irrigators and power sources for mills. River keepers were employed, banks maintained, fish reared for stocking, river weed cut, predators removed. The water meadows were kept in good shape not just for drowning but fishing as well. Suddenly the owners of the great estates began to value the rivers for the sport they could offer.

As the railways made the countryside more accessible, great houses hosted grand fishing parties. Gunsmiths turned their hands to fine reels, rods, lines, hooks and flies, using the latest techniques and materials. Weekly magazines like *The Field* and *Country Life* lionized innovators like Frederic M. Halford, a wealthy industrialist in his own right, who codified fly-fishing in a single book. Fly-fishing went from an obscure pastime to the 'must do' sport in a matter of decades. If you fished for salmon Scotland was the place to head for, but for brown trout dry fly-fishing the chalkstreams of southern England were the ultimate destination.

The mayfly period, or Duffers Fortnight, became as much a part of the English season as Ascot or Wimbledon. The future kings of England were elected president of the world's most exclusive fly-fishing club. Fine tackle manufacturers received the Royal Warrant. Government ministers cut short cabinet meetings to catch the train in time for the evening rise. Eisenhower took time out from the D-Day preparations to fish the River Test. As the fly-fishing craze spread across Europe and the Americas, visitors from abroad took home stories of the fabled chalkstreams which took on deserved iconic status. But time, money and enthusiasm are not always limitless, and as I walked around Gavelwood on this late September day I could chart the progression from a chalkstream paradise to something that is today a shadow of its former self.

Nobody set out to make it so. It was simply another twist in the evolution of the rural landscape. In succession the water meadows, watermills and finally fly-fishing were no longer part of the daily life of Gavelwood as the ownership changed to commercial farming. No longer were the myriad carriers and streams of any use, so they were left to atrophy. The meadows were ploughed, fertilized and sprayed for crops. The river was left untended. Gradually as the diverse habitat disappeared so did the creatures that inhabited the river, banks and meadows.

But three or four decades of neglect did not put Gavelwood beyond redemption.

3

WORK BEGINS

AS EVER, MY ancient leaking Land Rover provided little protection against the sideways rain of the late October day as I drove down the potholed track to the river. I had hoped for better weather. This was to be a landmark day in the restoration of Gavelwood: our first step towards the re-creation of something special, where the fruits of our dreams and expectations would, at least in part, be rewarded. After a month of back-breaking work North Stream was ready to be opened to fresh, gin-clear, chalkstream water from the Evitt for the first time in four generations.

North Stream is an ancient carrier that connects the main river – the Evitt – with another side stream we call Katherine's Brook. I say 'connects' in the loosest possible sense, because barely a drop of water has flowed through it in living memory. Along its entire length – about half a mile – it should really be a fast-flowing little river that takes the excess flow from the main river into Katherine's Brook, which in turn will rejoin the main river some 3 miles downstream. Instead the stream was a morass of fallen trees, roots, bushes, debris and mud.

I parked up close to the junction of the main river, where there is a set of hatches, built long ago, to control the flow of water into North Stream. Back in July, when we had first conceived the restoration plan, those hatches were almost invisible. On the river side a thick margin of reeds had choked what would have been the funnel-shaped entrance to the river. Today, the weeks of work had revealed three upright pillars of limestone, about the size of a tall man, set into the bank. They are slightly pockmarked in places, but generally washed smooth by centuries of water. The fronts of the pillars are V-shaped to deflect the current, and running down each inside edge is a groove into which are slotted oak boards – these regulate the amount of water that flows from the main river into North Stream. The oak is newly sawn, a lovely bright honey yellow that would, in a few months, turn to a silver grey. But for now their newness is proof that the hatches are repaired and ready to play their part in the rebirth of North Stream.

With everything Gavelwood has to offer – miles of main river, side streams and hundreds of acres of water meadows – North Stream might seem an unlikely candidate for the first step in the restoration. At first glance, if you noticed it at all, it looks marginal. It is not very wide – a reasonably agile person with a short run-up could leap it in most places – and is fairly straight, without any particular features that catch the eye. My suspicion is that given a few more years it would have disappeared entirely to become a soggy ribbon across a water meadow, its original purpose long forgotten. But the first time I saw it I knew it had the potential to become the most wonderful spawning stream for trout, salmon and maybe even grayling.

On that first visit as I walked down the bank, occasionally pushing aside the branches of the bushes and trees that choked the channel, a few small, bright pockets of gravel glinted back at me, lit by the rays of sunshine that cut through the gaps in the foliage; the gravel kept free from silt by the spring heads that bubbled up from deep below. Loose, well-oxygenated gravel is vital for spawning trout. It is the place the gravid female lays her eggs and the home for the ova as they metamorphose from eggs to tiny fry, out of sight from the many predators that see them as a nutritious food source. My hunch was that beneath the silt and overgrowth North Stream was a gravel haven and finding out was not going to be very difficult.

In fact it proved harder than I thought. A combination of wicked stinging nettles that are at their fiercest in the high summer, plus the barbs of the hawthorn and the clawing tendrils of the wild roses forced me back each time I tried to push my way down the bank. Eventually I came across an ash tree that had fallen across the river, flattening my access. Using a tree branch for support I slowly lowered one foot into the shallow water, letting my weight push it down through the thick mud, hoping that I would make contact with the riverbed before the water reached the top of my boots. Fortunately I did, and the firm base beneath my boots told me I had reached the best kind of rock bottom. I jiggled my feet and through the thick rubber soles I could feel the friable gravel. As I waded upstream I kicked away at the silt bottom to expose what I had hoped for – gravel the entire length of the stream. The further I waded up the more certain I became of the plan to have North Stream ready for autumn spawning – with clean, bright gravel where the trout eggs would be nurtured by a constant flow of fresh water from the main river. Yes, the timetable was tight and yes, the work would be hard, but at that moment to miss yet another year, after the decades of decline, seemed positively criminal.

I plotted the timetable as I walked. We needed to be finished by 1 November. River Evitt trout typically start the act of spawning around mid-December, but they would need at least a month to grow familiar with their new environment before beginning courtship. To have us clumping around would put an end to that before it even started. The trout fishing season ends on 30 September. It would be tempting to start clearing the stream earlier, but our downstream neighbours, not to mention our own anglers, who regularly fished at Gavelwood, would not thank me for sending muddy water and debris their way. So we had four weeks to take what looked like a clogged ditch and transform it into a piscatorial love nest and nursery.

There are two ways to restore a river: the easy but expensive and the cheaper but hard. The easy but expensive way involves signing up an ecological consultant who will start by carrying out a painstaking survey (at your cost) of the river and surrounding land. Every tree will be plotted, the curvature of each bend delineated and the depth of the pools plumbed. Soil and water samples will be analysed, flow rates monitored and the wildlife censused. In return for a mighty fee you will receive a mighty document with maps, drawings, graphs, commentary and appendices. You'll read it. Actually you won't – you will read the two-page executive summary at the front and glance through the rest. Fortunately your fee includes a presentation, so you head for the consultancy offices. Having been ushered into the boardroom by a receptionist you are then glad-handed by the team. Everything is very *exciting* and the possibilities *immense*. You can only agree, but how do I do it, you ask. At this point the meeting gets serious. Sitting across the table from you is the Chief Executive, who takes a copy of the report and places it squarely on the table in front of you.

'May I be frank with you, Mr Cooper?'

My advice to you at this point is to say no and leave; no good can ever come with a person who opens with this line. But you are curious, so you invite the man to continue. He opens by telling you what you know already. The report on the table is the perfect guide to do-it-yourself restoration. Everyone around the table knows this, but our wily Chief Executive casts a fly into your path he knows you will take.

'How much were you planning to spend on the project?' he asks innocently. You quote a number, faintly embarrassed that you thought it could be done for so little. He purses his lips. 'Here's the thing,' he says. 'You will do an OK job with that budget, but this is such a very *exciting* project, the potential so *immense*, that we should think big. Let's quadruple your budget, apply for funding, and in the end you'll only have to dip into your own pocket for a fraction of what you originally thought.'

The lure of his fly is too much and you rise to it like the greedy chap you are. The thought of twenty grand's worth of work for the cost of five is too much to resist. Leaving the room an hour later you have been truly hooked and landed. The consultants are delighted (but not surprised) with a new contract to seek out funding and manage the project when the grants roll in. You are of course still on the hook for their fees if the funding never shows up, but that is a discussion left for another day.

But I don't much like easy and expensive. It takes too long, the finished job is never as good, and it seems a bit immoral to me that half the money will go to consultants, however expert. And quite frankly, where is the fun in handing the project over to strangers? I wanted to get my hands dirty: stand in the river, look upstream and with a trout's-eye view of the world fine-tune the work as I went along.

But all this was still ahead of us when my team and I gathered in August to make plans for North Stream's restoration. It was not the best month to do our kind of survey – the undergrowth at its most dense, the flow almost non-existent – but we could see enough to make some educated guesses. The work was going to be done by Steve, Dan, myself and a team of irregular helpers.

Steve is the closest thing we have to a full-time river keeper. A retired fireman who looks forty but is in fact fifty-five, he runs triathlons just for the hell of it. He can, and does, work all day felling trees, cutting weed and hammering in fence posts. He is in fact more of a coarse angler, and Gavelwood sort of inherited him when some local lakes closed down.

Dan is young. We tease him for being young and he mocks us for being old. In his early twenties, Dan is on a sabbatical year from his university ecology course. I have a feeling he may have dropped out for good, but it is a suspicion I have kept to myself.

The irregulars are a band of loyal fishermen and locals who simply like to help. They turn up as they wish, or Steve will put out a call when he needs some extra hands. It seems to work and every few months I put some cash behind the bar at the pub for an evening of merriment. Work on the river next day is sparsely attended.

On that particular August morning Steve, Dan and I had gathered at Bailey Bridge, a steel latticework bridge of the same name that was invented by the British army. You used to see them all over the river valleys at one time, but most have rotted and rusted away. Built of light steel and wood, in sections small enough to be lifted into place by hand, they were ideal for bridging meadow streams. Designed to take the weight of a tank, they were much loved by farmers, not least because they were easy to 'liberate' from the nearby military camps on Salisbury Plain if you drank with a friendly sergeant major.

Our bridge looked to me like it was getting towards the end of its life, but we estimated that by replacing a few of the wooden boards and repainting the metalwork we could eke a few more years out of it. I had my doubts about its inherent strength but Steve was prepared to test it out by the simple act of driving a tractor and laden trailer over it. Sometimes he worries me.

The first decision we needed to make was whether to clear one or both banks along North Stream. Both sides were equally overgrown, and there are merits whichever way you choose to go. In sheer practical terms opting for a single-bank restoration halves not just the work required for the initial clearance but also regular maintenance in the years to come. With our tight timetable it was an attractive proposition, but ultimately we had to decide on what was best for the wildlife, the river and the fishing.

Stepping off Bailcy Bridge and towards the stream, our path was blocked by chest-high stinging nettles. Nettles are no great friends of ours – sure, they are much loved by caterpillars, who feed voraciously on them, but for the river keeper and angler they are a menace. They grow fast, crowd out more useful bankside plants and sting like crazy. Fortunately getting rid of them is not hard, at least if you have someone like Dan to do the work. Nettles are nitrogen addicts – in their effort to run wild they suck every last drop of nutrient out of the ground. But when they die back in the autumn the rotting stems and leaves put nitrogen back into the soil ready for next year. However, cut the nettles down and rake away the cuttings and you deprive the next generation of their nitrogen fix. Other species soon encroach on the ground left bare and new plants thrive in place of the nettles. For Dan a couple of weeks with a scythe and rake were on the cards.

Beyond the nettles and bordering the stream was the scrubby woodland that ran the length of North Stream. On both banks it was 10–15 yards wide, because some years earlier it had been fenced off. The fence was pretty much all but gone, save for a few posts and rusting strands of barbed wire that would no doubt trip us up at some point. The main growth was really stunted hawthorn, which had done us something of a favour in the absence of the fence, by keeping the cattle away from the banks and out of the river. Pretty in its own way, and home to the hawthorn fly, we mulled over how many of these bushes-cum-trees should stay, be trimmed or cut down. I am a huge fan of hawthorn. It is the constituent element of every hedge in the chalk valleys and in April its vivid lime-green leaves and white or red flowers are the first tangible proof of spring's arrival. Admittedly the flowering bushes do emit the most awful stench, which makes you think there is a rotting corpse under every hedgerow, but once you know what it is it does not seem that bad.

What's more, the hawthorn fly or St Mark's fly (*Bibio marci*, so called because it hatches around St Mark's Day on 25 April) causes much excitement among fly-fishermen in the first few weeks of the season, not least because trout go on quite the feeding frenzy when these clumsy fliers drop onto the river surface. The fly has no real connection with the river, so why trout go mad for these freakish-looking creatures is a mystery about which one can only hazard a guess. At first glance the hawthorn fly looks like an athletic housefly, but at second you'll see its long spindly legs dangling below it, like the undercarriage of an aircraft, with big knuckles for knees and so hairy you might even stroke them. The flies don't live for long, maybe a week at most, having emerged from larvae in the soil beneath the hawthorn bushes. Once hatched they hug the hedgerows for protection from the wind, but from time to time an unexpected gust will whisk them across the meadows. From this point on things get tricky. They are, without shelter, the most hopeless fliers and you will see them buffeted by the breeze. Occasionally when the wind drops they regain control, but it will be short-lived and once over water they will plop onto the surface. Unable to break free of the surface tension they are easy pickings for the trout.

Along the length of North Stream and among the hawthorn are a few spindly ash, plus some alders, clumps of hazel, brambles and wild roses. The trees we wanted to keep we marked green, those we would thin, blue, and the rest – marked red – were to be cleared. It soon grew abundantly obvious that on this bank there was not much to preserve, whilst on the opposite side pretty well everything, bar a few branches that were falling into the river, could remain undisturbed as a sanctuary for the creatures that live along the riverbank.

Part of the restoration process is about letting light back into the river and onto the riverbed itself so that the weed there can grow. The term weed does these river plants like crowfoot, starwort and water celery something of a disservice. Weed implies that they are invasive and bad, but the reverse is true. The right river weed, in the right river, is home to nymphs, snails and all manner of tiny aquatic creatures. It provides cover for fish, shade from the sun and refuge from predators. And as a filter for the water, a healthy river needs healthy weed, and that will only grow with sunlight. It is hard to say anything bad about weed, and a chalkstream without it is on a downward spiral.

Removing a fair amount of the thicket growth along the south-facing bank was going to suit us very well. In this respect clearing the north bank alone would not have helped, because as the sun tracks east to west across the sky during the day it would have left the stream perpetually in shade. If you ever doubt how bad perpetual darkness is for the ecosystem of a river, glance under a bridge one day; it will be as bleak as the surface of the moon. That said, our work was far from about eliminating all shade; trout and all the creatures thrive best where there is a mix of light and dappled shade, so before we took the saw to any bush or tree we cocked our heads to each in turn to decide stay, trim or go.

All the way up North Stream the stream itself was no great issue for us. Sure there were plenty of branches and stumps to pull out, but the dark shade had pretty well prevented anything growing. Once the obstructions were removed the sheer volume of water over the winter would flush away the mud and slime. That was of course always assuming we were able to open up the Portland hatches.

Removing the decades of compacted silt could be done by hand but it would be long and laborious, so we elected to bring in a digger to do the job. Machines are great, but sometimes you have to go easy with them or risk doing damage to the very things you wish to preserve. The Portland hatches were a case in point. They had stood the test of around 500 years because they had been carefully constructed with strong foundations. Smash those with the digger bucket and our problems would multiply.

Steve produced a steel rod with a T-bar handle. Jumping down onto the silt he pushed the rod into the ground until at around 5 foot down we heard a muffled clunk. He tapped the rod up and down twice to confirm that he had hit something solid. Over the next hour, working like an avalanche rescue team on snow, we each took a rod, gradually mapping out the depth and extent of the stone slabs ready for the digger to do the hard graft once the season had closed.

There is never what I would call a really good time to embark on a restoration; every month, every time of year has its merits, but inevitably there is disruption to the natural order of things – removing the bad and encouraging the good. The bad comes in all shapes and sizes: people, fish, animals, mammals and even plants. Yes, there are even bad plants on the chalkstreams, the most invidious of which has its origins on the foothills of the Himalayas.

My problem with Himalayan balsam is that I rather like it. The tall plants stand high above the surrounding vegetation in vast swathes and the light red-pink funnel flowers are a sea of colour that gently waves in the late summer breeze. The smell from the flowers envelops the riverbank. It is a dry, sweet smell – lightly medicinal and cathartic at the same time. It is completely alien to anything else that grows in the meadows. It looks different, smells different and has the most amazing way of distributing seeds when the flowers have died and the tall plants are denuded of leaves, just leaving brown seed pods. Brush past the balsam and the seed pods burst with an audible 'pop', shooting their kernels yards around. It happens with quite some force; you will feel the sting if they bounce off your face or hands. Young children love to grasp the plant at the base, shaking it with all their might while the rat-a-tat of seeds sails harmlessly above them.

Imported as an exotic species from Nepal in the early 1800s, Himalayan balsam is now established in Britain, but has had particular success on rivers where the seeds, which can survive two years, are distributed by the water. As a single plant it is no great problem, but that is not in the nature of Himalayan balsam. It is an invader that grows faster than any native plants, shading out and eventually killing all others. Walk the banks in October where the balsam has taken hold, and the area looks like a wasteland. Everything beneath the balsam is dead. In truth it looks like the ground has been sprayed with a toxic weedkiller, and come winter, that soil is bare and ripe to be washed into the river.

Fortunately for me and river keepers everywhere, Himalayan balsam is an enemy that can be defeated. For now, in October, there is not much I can do, but come early summer when the balsam pops its heads above the surrounding growth we will walk through the meadows pulling out the plants by hand. Mercifully they are shallow-rooted, so they come out easily or snap off at the base like soggy celery. However, not all my enemies are so easily defeated.

Mink have thrived in the abandoned Gavelwood. Wily creatures, the thick undergrowth and clogged streams are heaven-sent for this predatory invader. Predatory they certainly are. Fish, water voles, field mice, duck chicks, frogs, baby moorhens, even rabbits – if it moves mink will eat it. The mink I see at Gavelwood are American mink – *Neovison vison* – which first arrived about a century ago as escapees from the fur farms that were established between the world wars. Despite their fearsome reputation they are really quite cute; I always think they look like a bigger, elongated version of their favourite prey, the water vole. Mink have the most beautiful dark brown fur, almost black in some light. Strangely, though all the mink you spot today are this colour, they all originated from the light-coloured mink imported by the fur trade. Clearly, however, being white in green meadows was a poor lifestyle choice. After a century of wild living it is a moot point as to whether mink can still be regarded as an invader. Non-native definitely; indigenous never; but successful settlers yes. They took hold at precisely the same time that the otters declined. It was no fault of the mink that the otter almost became extinct in Britain, but nature abhors a vacuum.

On that filthy late October day the success of our survey and the work Steve had done with the digger was there to see. The silt and mud were gone, smeared over the grassland around the hatches. The wet surface of the slabs on the base of the river glinted back at me. Some of them were truly huge; a full 10 feet square and nearly a foot thick. One could only wonder at how they were ever put into position all those centuries ago. The digger stood by ready to drag out the reeds within the hour.

I wasn't exactly sure where Steve, Dan and the irregulars were working that morning, but the whine of the chainsaw through the rain from somewhere far downstream gave me a rough idea, so I followed the noise. For all our hard work over the past month North Stream really looked in quite a sorry state. I am tempted to say worse than when we had started, but it is always this way, a sort of darkness before the dawn.

The ground along the bank was churned up; deep ruts showed where the tractor had strained and dug deep to pull out the worst of the trees. Every so often I would come across a round circle of ash where the lads had lit fires to burn up the detritus. There was a pile of tree stumps, too big to burn and unwieldy to cut up, so they would be taken away to be dumped and end their lives in a rotting heap. This would be a paradise for woodpeckers seeking easy food and a palatial home for woodlice. From time to time I came across some long, straight tree limbs which had been carefully trimmed and set aside. This was our kind of recycling; logs and branches that would be useful for building weirs, flow deflectors and groynes in the river when we reached the next stage of the restoration.

The entire bank was pockmarked by tree stumps, cut level with the ground: the cream white of the ash; dark red of the alder and burnt orange of the hawthorn. The hazels looked like bundles of cigarette filters pushed into the ground. The fact is we only pulled out the stumps we had to; by far the most were left in the ground. There is no point in ripping them out, as the root structure will live for years and bind the bank together. Some of the stumps will sprout again, indeed species like the alder thrive as a result of the extreme pruning. And as North Stream evolves over the coming years, we'll let some grow back into mature trees for cover, shade or simply extra interest.

If I thought the banks looked bad, the stream itself looked worse. The water reflected the sky; it was dark and gloomy. Barely flowing, the surface was covered in twigs, chainsaw shavings, dead leaves and chopped vegetation. On the far bank the bushes and trees had shed all their leaves, the spindly branches dripping from the rain. The only comfort I took was in the windbreak they provided from the north wind whipping the rain across the meadows. A north wind is the enemy of all fly-fishers – the cold kills off hatch and stops the fish feeding, which gives rise to that old saw, 'When the wind is from the north only the foolish angler sets forth.' However, today was not the day to worry about the north wind, which is an almost daily occurrence in winter, sweeping as it does down the river valley. Days like this are always the fun parts of a restoration, when months of planning and weeks of work come to fruition. Today all we needed to do was dig out the plug of reeds, lift the boards in the Portland hatches and let the river flow in. And that is what we did.

Steve used the digger to scoop out the reeds and we laid them to one side. The flag irises, with their yellow flowers that bloom in May and June, were too beautiful to discard, so we'd replant the rhizome roots down North Stream to kick-start the regrowth in the spring in the parts of the stream left bare by their removal. The reeds gone, the water started to build up against the honey-yellow oak boards. These boards are never completely watertight; water weeps through the gaps between them. So as the flow backed up and the pressure increased, water squirted through the holes as if from a hosepipe. We were ready to open up. Standing on the bridge boards over the hatches we worked in pairs to lift the top boards out. The lower four quickly followed and within moments the fast flow from the Evitt rushed into North Stream. Like excited schoolboys we followed the bulge of water as it forced its way downstream. From time to time it came across an obstruction. Then the water would begin to back up, but when the force grew too strong the obstruction would give way and the flow continued. On it went down North Stream, carrying the mud and debris in its path. Under Bailey Bridge and the final straight to the main river. Standing by the deflector we watched the confluence of the two currents, the first time anyone had seen this for maybe forty or fifty years. True it wasn't the prettiest of sights, with the dirty water of North Stream adding a nasty stain to the clarity of the Evitt, but the knowledge that our plan had worked was enough for now. Given a few weeks North Stream would flush itself clean, and then the fish would return.

4

SPAWNING AND THE CYCLE OF LIFE

AFTER THE FRENETIC activity of summer I miss my riverside companions on a winter dawn morning. No reed-chewing water voles suspiciously eyeing my progress along the riverbank, plopping for safety under the water if I come too close. No dew-laden spider webs strung between the purple loosestrife, glinting in the rising sun, as an eager arachnid crabs with intent across the translucent filament harvesting the victims of the night. Even the rabbits have gone, and as for the lolloping hares, no chance of any of those until spring. But even if it is all quiet along the banks, in the ever-clear water of the chalkstream the game is on to create the next generation of trout and salmon.

Trout and salmon are often spoken of in the same breath, but they are in many respects as close to each other in genetic terms as a horse is to a zebra. For a fly-fisherman they define what you are on a river. As salmon and trout are two distinct breeds, so are the men that fish for them. Not to announce which you are, even though you might fish for both, is like saying you support the Manchester football team. United or City? Salmon or trout? Both are equally tribal.

For fish whose subsequent lives will diverge so totally they begin life in the same gravel beds, of the same rivers, at precisely the same time of year. In lives that will span five to seven years some brown trout will travel no more than a few hundred yards from their birthplace, whereas the salmon has a round trip of some 4,000 miles to complete its life cycle. While we may think of a salmon as a river fish, in fact the greatest proportion of its life is spent at sea. These salmon are Atlantic salmon – *Salmo salar*. Defined as anadromous, their natural habitat is the sea, but they must return to the river of their birth to spawn. The eggs are laid in a river and that first year of life, as they grow from fry to parr and then smolt, is all spent in fresh water. But no chalkstream could ever provide enough food for a salmon to grow to maturity, so at a year old, measuring no more than 6 inches long, they head for the ocean and the food-rich waters off Greenland. It is an epic journey that begins and ends in a stream no more than 15 yards wide and a few feet deep.

The spawning grounds created by salmon and trout in the gravel riverbed are known as redds, and the sight of the first redds, be it in October or November, is something of a red-letter day for us chalkstream watchers. Indeed, redd-spotting becomes something of an obsession from around October time. I say 'around' because rivers don't obey the Gregorian calendar. Like the snowdrops in your garden that appear in January one year and February the next, the creatures of the river adapt their habits according to what's happening around them, which is in turn dictated by the climate. And not only the weather of now; the effects of a dry summer or harsh winter for instance, may linger many months or years to come.

I'll get excited text messages from river keepers: *Seen a redd today. First of the year!!!!!!!!!!* ☺. It is exciting because amid the gloom of late autumn and the winding down of a fishing season, it is a small ray of hope for things to come, however distant. It is also proof that as a river keeper you are doing something right. Your river is so damn perfect that fish want to breed in it. How good is that?

Walking beside the river, you will find the redds are easy to spot once you know what you are looking for: pale lozenge-shaped indentations on the river bottom, with a mound of gravel at the downstream end. Brushed clean of silt and debris, they shine out like lights compared with the surrounding gravel. Sometimes there is just one, sometimes a cluster, but it is the size that immediately marks out the difference between a trout and a salmon redd: the former about the size of a snowshoe, the latter a good-sized door mat. And there is more latent intent about the salmon redd; it will be dug deeper, down to the hard base beneath the gravel. Random gravel stones from the digging will be scattered far and wide across the riverbed. The mound of stones at the end will be much higher and more pronounced. Redds are, of course, made by the fish themselves to harness the flow of well-oxygenated water through the loose gravel to incubate their eggs, and oftentimes the hen will lay in more than one redd. Laying in a single redd is quite literally putting your eggs in one basket, and that basic instinct to perpetuate the species drives the hen to hedge her bets by laying in a series of redds, maybe with other hens. But we have to track back in time to appreciate how and why we have arrived at this point.

From my daily walks up and down the river the progress of the trout from an everyday *Salmo trutta* to a body quivering as if electrocuted whilst he releases his milt over the eggs is far easier to track than that of *Salmo salar*, the Atlantic salmon. It is in September that I start to see the first signs of spawning in the trout, who start to change in appearance in the weeks before they start the actual process of cutting redds and spawning. Suddenly that headlong pursuit of every item of food to feed on in preparation for the winter ahead slackens off. The fish are just as active, but not for food. Somewhere in their fishy brain the search for food is replaced by the search for a mate. The change sweeps over their body and suddenly that golden-brown complexion is replaced by a fierce red blush along both flanks. The males sprout a vicious-looking hook – a kype – on their jaw. The kype is largely for show, but it does make an otherwise innocuous-looking trout look like someone you would not want to mess with.

During this time the salmon are absent from this river, still making their way along the English Channel from the Atlantic to pick up the scent of their birth river somewhere on the south coast. How salmon navigate the entire journey to the far side of the Atlantic to the waters off Greenland and back again remains something of a mystery. The position of the sun, the stars and the gravitational pull of the earth are all cited as guides, but it is certain that the final leg of the journey is determined by smell.

Salmon never look to me like creatures that depend on smell for survival – their incredible ability to leap huge waterfalls or swim unceasingly for months on end seem more important – but smell is the thing. Early on in their lives they imprint the odour of their birth river onto a hormone that is secreted in the thyroid gland; it stays with them for evermore. Their hormonal library of smells is highly selective; only the ones that really matter make it onto the data bank. Likewise they will log the odour of their brothers and sisters in the river, picking up their scent in later years when the shoals are travelling across the ocean.

By the time our salmon sniffs the first scent of home, he or she has surmounted incredible odds to make it thus far. Of those 5,000 eggs laid three years ago in the River Evitt, our *salar* is probably the sole survivor, or at best one of two. And the dangers are far from over. Ravenous seals are gathering for an autumn feast and the drift nets in the estuary are laid in wait. It is the misfortune of salmon that they make such good eating, though it should be of no surprise. They are super-fit and have spent the past two to three years in the beautifully clean water of the Greenland Sea eating nothing but squid, shrimp, crustaceans, small cod and mackerel.

As far back as medieval times salmon has commanded a premium price, so the ever-resourceful coastal communities around Britain developed the highly efficient drift net to capture the salmon returning from the sea. There are all manner of types of drift netting, each of which has evolved for the particular locality, but the principle holds good for them all: wait for the tide to go out and then set your nets in such a way that they intercept the salmon travelling towards the estuary bottleneck on the inbound tide.

Travelling around the coastline of Britain you will see all sorts of weird and wonderful nets rigged up to capture salmon, though they are becoming fewer. Declining runs of salmon, fierce campaigning by conservation groups to have the nets removed and the harsh demands of a truly hard and difficult job are all contributing to the decline.

The simplest form of drift netting is a long net, anything from 30 yards to a few hundred, and 5 to 10 feet deep, that is slung across the tide, supported by floats along the upper edge. In shallow water it will be held in place at each end by a man holding a pole; in deeper water by boats. As the tide races through, the salmon follow, to get caught in the mesh of the net. The netsmen gather the ends of the net into a circle, capturing the salmon by hand as the circle gets smaller.

Not surprisingly this fast, efficient method is the one most favoured by poachers, but it is the fixed nets that are more typical of traditional salmon netting in the estuaries: wooden posts supporting nets that face the incoming tide. Sometimes the nets will be shaped like giant boxes as large as a van, 10 to 15 feet above the beach level, open on one side. Other times they are funnel-shaped. Or most simply, nets loosely slung between posts like a garden fence. One way or another they are doing the same job of entangling the salmon, which become more trapped the more they struggle. As the tide ebbs some of the fish will escape, but once the nets are exposed to the air the salmon don't have long to last and the netsmen will appear to complete the harvest.

Fortunately for our salmon heading for the Evitt, these dangers are slight. The seal population along the south coast is sparse compared with say the northwest of Scotland and the estuary netting is now less common than it once was. Ahead is the brackish water of the estuary and beyond that the purity of the chalkstream water, although this change from seawater to fresh will quickly change our salmon's physiology as the body cells, body and organs adapt for the months to come. Such problems do not trouble our brown trout, whose sole mission at this point is to find a mate. No swimming over thousands of miles for him. His potential partner may be one of the other trout that have lived within a few hundred yards of him for all their lives. There never seems to me any great logic or grand plan to the way trout choose their partners. There will be a bit of swimming around, occasionally another male will sidle up beside a paired female to be promptly chased away, but on the whole it all seems to happen at random. Or that is how it looks to me with my bank-down view, but scientists think there is more to it than this, and that even fish have that 'eyes across a crowded room' moment when the right mate comes into view. Nobody knows exactly what is going on in and around those redds, but somehow skin colour, conformation, size, pheromones or possibly a mix of these and other factors combine to make their choice a complex matter.

You would think that the trout at Gavelwood would get used to me – after all, they must see me just about every day. But they never seem to. The only time they are oblivious to my presence is at spawning. Every other time they will bolt for cover if I disturb them. But at spawning I can stand on the bank almost directly above them and wave my arms about like a lunatic and they will carry on with their business. If I am in the river I can damn nearly step over them in my waders as they fin away one side or another to let me pass, to return to the same spot in my wake. The creation of the redd and the spawning to follow is all-consuming, but in their enthusiasm for each other the trout also forget about their most dangerous predator, the otter.

Otters are an ever-present but rarely glimpsed part of the Gavelwood family, about whom I have mixed feelings. On the one hand I should celebrate their existence, having come back in great numbers from the brink of extinction. But on the other they are one of the best fish-eating machines invented by Mother Nature. In ten months out of twelve it will be rare for me to see an otter, though most days I will be able to tell they have passed through. They are nocturnal creatures, using the river as their highway, travelling as much as 20 or 30 miles in a night. A paw print in a muddy bank, crushed grass where they slide into the river, and spraints, or dung, are all telltale signs, as is the corpse of a fish. In winter the latter will be barely recognizable as a fish: a few bits of fin, skin and scale in an area of flattened grass where the otter will have settled down to eat. If it is, or was, a gravid female trout there will be a scattering of eggs to add to the mix of body parts, which is sad in its own way, though they will rarely go to waste as the moorhens and water voles, or maybe a passing fox, are more than happy to feast on this unexpected bounty.

Every time I see the dog otter – I think we must have just the one – I am struck by his size and lithe movement. He is around twenty pounds in weight; put in context that is about twice the weight of a domestic cat. And like a cat he is incredibly supple; he doesn't dive or jump into the river, he pours himself in, barely making a ripple. Once in the water, despite his bulk, he is hard to spot, but I'll be able to track his underwater progress by the surge he creates on the surface. Some yards downstream he will pop his head out of the water, swivel around to check he has put sufficient distance between us and continue on his way at a more leisurely pace.

It is that bulk that makes the otter a deadly predator, because the bulk requires constant nutrition. It is reckoned that an adult male has to consume 10 per cent of his body weight each day to survive the winter. That is a two-pound fish, which is a big fish for the Evitt. More realistically we are talking about a whole bunch of smaller fish from trout, grayling and eels right down to the tiniest, like bullheads. Of course my otter's diet is not pescatarian – frogs, crayfish, birds and water voles are all fair game – but in winter there are meagre pickings, so a careless spawning trout is a tempting prospect.

Come the summer things are very different; food abounds and so do the otters, which often become my evening companions when I stay late to fish the evening rise. I always hear them before I see them. Otters have this high-pitched 'eek' noise that they ping across the meadows like sonar to keep in touch with each other; it is the mother's way of tracking the pups. As night falls the parents seem perfectly content to let the young ones range all over Gavelwood. I can become caught in this crossfire of constant eeking, and it is not a noise you have to strain to hear. It is incredibly insistent and frequent, though the frequency is a good indicator of how well things are going. I reckon that a contented otter eeks every thirty seconds; if one becomes distressed the frequency escalates until it becomes almost continuous. And it lasts all night or until the family move on to another part of the river.

These long, light summer nights are part of the growing-up phase for the pups, when the parents bring them out from the birthing holt to learn how to explore and hunt. Strangely, unlike most other species that become fiercely protective of their young, otters are more playful. They will often swim and hunt together in the river, just keeping a weather eye out for me. Otters are pretty well the top of the food chain and they regard humans as more of an oddity than a threat. However, lower down the chain the poor fish truly suffer.

Two fit adult otters, plus three or four ravenous, growing pups, seem to be the usual summer contingent that I will see in the twilight and on into the early hours of the morning. The solitary otter I see in the winter is a stealthy hunter, but in the summer the pack instinct takes over. Pike Pool, about halfway down the main river, is our deepest part of the Evitt and is the favourite place for the family to gather for a hunting lesson.

The pool, which starts when the river makes an abrupt 90-degree turn, goes down to about 15 feet, constantly eroded by the water as it hits the opposite bank and swirls in back eddies before the gradient reasserts the natural order of things and the water heads downstream as it should. Along the bank stands a line of alder trees and the roots grow down into the water. Beneath the roots is a huge undercut, the perfect refuge for the fish and eels, or so they think.

The family will not so much hunt in a pack, but they do hunt collectively – rapidly diving and surfacing across the pool with their wet, brown fur glinting in the moonlight. They pause for just a moment to catch their breath before diving again. One can only imagine the massive panic in the fish community as they flee for safety in the dark recesses under the tree roots. And safe it is from every predator other than the otter. For herons and cormorants the fish are protected once out of sight. For mink they are too deep. Pike usually give up after a single attack. But otters are persistent. Once they have the fish cornered, they will dive and dive again. As the hunt becomes more frantic and the effort greater, they will emit a sharp cough when they surface to grab a breath. Inevitably they succeed and the victorious otters will slither out of the water onto the base of the tree to start devouring their catch. They sit back on their haunches, holding the fish in front of them using the sharp claws of their webbed feet for purchase, and then tear at the body, starting with the head. It is violent and fast. From the other side of the river I can hear the flesh being torn apart. Strangely they are not competitive about the catch; they wait their turn. When one has had enough he or she will lay what is left down for another to pick it up.

I have never yet seen the otters catch a salmon; maybe they are too big or simply swim away fast rather than hide. Trout are the most common, eels not far behind, and grayling the most prized – in winter they devour every last morsel of the latter. In the summer part-eaten fish or eels, too big for the otter pups to finish, are common. With the eels the head seems to be the only bit they like to eat; decapitated eels are a common sight in the morning dew. I usually kick them back into the river for the crayfish. I used to throw the part-eaten fish into the field – dead fish on the riverbed can look alarming to visitors – but since I have discovered that otters are partial to a five-day-old, decomposed trout I also kick them back in on the grounds that it might save the life of another fish.

It is something of a fallacy that trout love the fastest water in a section of a river to live out their lives; in fact almost the reverse is true. The older and bigger a trout becomes, the more he or she gravitates to the deeper, slower parts, so autumn is the only time we get to have a good look at the long-term residents who are the brood stock for the next generation. If you are a tiny little juvenile trout the fast, shallow water is a great place to grow up because you have the place to yourself.

For the bigger trout the effort of holding station in the riffles, the fast-flowing shallow water that separates the pools, is too much for any possible rewards and the risk from predators like herons very high. But for the little, tiny trout even a good-sized pebble will provide shelter from the flow whilst waiting for a tasty nymph to come tumbling by. Predators? Well, when you are small it is all about the lesser of evils. Yes, you could be plucked from the stream by a kingfisher, but in truth your greatest danger lies from the very adult trout that probably spawned you. The one thing all fish love to eat is other fish.

The trout I hoped would gather on the gravel beds in North Stream would be fast developers to do so at three years; four is more common and it is the females who first seek out the ideal patch to set up the nursery. It is true that fish often head upstream to spawn to seek out the purest water and best laying gravel, but unlike say Pacific salmon that congregate in the uppermost point of a river in a giant, swirling pink mass, brown trout are smarter than that. Quite frankly they travel only as far as they need to travel, be it a metre or a mile, which is why I had high hopes for our newly restored stream. Brown trout are eminently practical when it comes to spawning; if they have to travel 20 miles upstream to find the perfect place and mate they will do it, but if both are within a few yards, why bother? I was hoping North Stream would be that place, the breeding ground for the trout that inhabited Gavelwood already.

The main river was fine, but the stream would be better with more places for redds and a better nursery for the eggs once hatched. From my point of view, it was all about making it easy for the female, because creating a redd is tough work. She positions herself over the chosen spot and then with flicks of the tail or a sideways movement of the body gradually dislodges a few pieces of gravel at a time. With thousands of movements, executed thousands of times over a period of days, gradually an indentation is cut in the gravel of the riverbed. Some of the stones get carried away on the current, but others gradually pile up in a mound at the downstream end of the cut. This mound, seemingly an unimportant by-product of the excavation, will in fact be vitally important when the females come to lay their eggs. But for now our female has to seek out the right location for her redd. The main river is just too fast in most places, as no sooner will she start to dig a hole than the rapid flow will scour it flat again, and even if she succeeded, when it comes to mating the eggs would be whipped away in the current before fertilization had had a chance to take place. So in the search for the ideal spot I am hoping that the trout moving upstream will turn right into the relative calm of North Stream to check it out.

Every action in a river causes some sort of reaction, so digging up the riverbed, however well intentioned, causes all sorts of commotion for other river creatures, and in this particular case the tiny ones. The gravel riverbed is home to millions of invertebrates, animals like snails, bloodworms, nymphs and shrimps, which thrive in the constant temperature of the chalkstream water. While 10°C might be a very cold bath for humans, for this group it is perfect. And if they thrive, so do the creatures that eat them, namely the fish. Fish are opportunists. Unlike people they don't have a routine that tells them it will be lunch at such and such a time. If food comes along they eat it and the moment that the redd cutting begins I will see the yearlings – fish under twelve months old – gathering below the cutting area to start hoovering up the unfortunate invertebrates, who can only drift helpless on the current until they either get caught in some weed, float down to the bottom or get swallowed. It must also be said that the yearlings, or parr, are not just there for the food; as eager adolescents they are standing by to add their bit to the spawning process. These 'sneakers' as they are called will slip between the adults at the crucial moment. Whether they contribute much in a normal year is debatable, but nature brings them to sexual maturity early as a back-up plan. In a bad year, maybe caused by low water or some other natural disaster that prevents enough males making it to the redds, there will at least be someone there to complete the job.

Fish are not beyond digging into the gravel themselves to find food. Watch a grayling in a river and you will see him go tail up, push his snout down into the gravel and with a puff of silt around his head suck up a shrimp. But why go to all that effort when a redd-making trout does the work for you? This is a winter feast that will only be bettered by the trout eggs themselves. And in the hot summer days, when anglers start to feel the heat and the fish get lazy, there are opportunities for both to capitalize on the dislodged food sites. At four or five spots across Gavelwood water meadows I have places where the cattle can either wade across the river or get into it to drink. As your average bovine drinks around seven gallons a day, maybe twice as much in hot weather, that is a lot of getting in and out of the river. And every time they do it stirs up the riverbed, uprooting the inhabitants. Trout get to know this, so they wait downstream, only moving out from the shade when the muddied water gives them notice of food to come. I do the same, and a well-cast shrimp imitation as the clouded water starts to clear will often turn a dead afternoon into a successful one.

The gravel of North Stream was abundant, but the decades of neglect had left it rock-hard, without the winter floods to break up the surface and sweep away the silt that had formed a crust. Within a week of reopening the Stream the worst of the silt and mud had been washed away to reveal plenty of potential spawning grounds, but when I tested them out the reality was depressing. Jabbing a garden fork into random sections of the riverbed I was mostly rewarded with a bruised hand. The tines would barely penetrate more than an inch or two. This was bad news. If I could not break through with a steel fork then the trout would find the same and keep moving on upstream to abandon North Stream. There were two options – do nothing or intervene.

Do nothing is not so bad if you don't mind waiting for years. Gradually, nature, in the form of exceptionally heavy winter flows, would break up the surface into the loose gravel that a trout might easily dislodge. But I didn't feel inclined to wait for years, so intervention, in the form of gravel-blasting, was the remedy. Gravel-blasting is not the nuclear solution it might at first sound. You take a high-pressure water pump with a steel probe on the end, stand yourself in the river, press the probe down into the gravel to a depth of about 6 inches and then wait while the water from the pump does the work, washing away the decades of silt that was binding together the gravel stones. When the water starts to run clear, you pull the probe out and push it back into the gravel a foot or so away. For the first ten minutes this is a fun job, but after a while the novelty palls. It is effective, however, and when you stand on the bank to admire your handiwork there is always a certain amount of satisfaction – the riverbed looks like a freshly plumped pillow and the gravel will positively glisten.

While the trout are starting to weigh up the options of North Stream, our salmon pick up the pace as the scent of the home river gets stronger. Past Land's End they start to hug the coastline, the beaches of Cornwall then Devon almost in sight. Gradually the pack thins out as one by one they peel off for rivers like the Dart,

Exe and Camel. For the remainder the chalk cliffs of the Jurassic Coast are the marker that these salmon are heading for the chalkstream rivers of Dorset, Wiltshire and Hampshire. Yet the salmon might be less eager to make the transition from salt water to fresh if they knew that the change signals the end for most of them: nineteen out of every twenty salmon are certain to be dead within a few months. The odds are much worse for the males than females, but they know nothing of this, so the urge to procreate impels them forward.

Why do they die? From the very moment our salmon enters the fresh water he or she stops feeding. And this cessation is absolute. Not a single calorie of nutrition will be consumed until the salmon returns to the sea or more probably dies. During the time of its life when food is most needed there is none. Day after day the fish swims upstream against the current, navigating weirs, dams and obstacles whilst the body adapts to the change from salt to fresh water, losing weight and condition. In the confines of the river a new raft of predators awaits; otters, pike, herons, cormorants and even fishermen line up for a piece of the action. These are not good odds, and at the head of the river the body-sapping ritual of mating will deliver the death blow to nearly all who make it that far. At Gavelwood we are about 35 miles up from the coast, more or less two-thirds of the way up the river system. For a salmon, a chalkstream like our Evitt is an easy run: no massive waterfalls to leap over or fierce currents to swim against. The greatest point of difficulty is Middle Mill, 5 miles up from the sea. At one time this was the biggest flour-grinding mill in the county, capturing the entire river flow to drive two enormous waterwheels. Below the mill races is the mill pool, a huge expanse of swirling water, which is the first place the salmon rest up on their run inland.

When they first arrive from the sea, salmon are bright, bright silver, or to use the jargon, 'sea-run', the males and females almost indistinguishable from each other. If you capture one at this point, or get close enough to look, you will see it has sea lice attached to its body. For wild salmon these naturally occurring parasites attach themselves to the body while at sea and look a little like a tiny tadpole. The salmon receive no benefit from these unwanted fellow travellers, but fortunately having left the salt water the lice soon die and drop off.

At Middle Mill salmon used to leave the pool via one of the six relief hatches at the side. This involves a leap to take them from the pool, through the hatch and up into the river above to continue the journey upstream. The hatches are not wide, at most 4 feet, and at one time the miller used to hang hessian sacks stuffed with straw to either side of them, to protect the salmon that misjudged the leap and hit the hatch walls. Today, with the mill derelict and the hatches untended, there is a salmon ladder to do the job, a series of man-made pools that the fish gradually ascend one at a time. Once through the ladder it is a free run to Gavelwood.

It is around Guy Fawkes Night that I start to keep a lookout for salmon. By now the trout are full at it on making their redds, but the salmon will gradually appear one by one. The fish look different to when I had seen them at Middle Mill a few weeks earlier. The silver has gone and I can now tell a male from a female at a glance. The male has turned brown – you might almost mistake him for a giant trout – the kype has started to sprout and he will weigh anything from five to twenty pounds, with the very occasional one weighing thirty pounds or more. He will sit immobile on the riverbed for hours or days on end. Sometimes I have to strain to work out whether it is a fish, log or lump of weed. Occasionally a new arrival will make a few leaps in Pike Pool, but on the whole they arrive like submarines on silent running.

The female doesn't change as much as the male; she tarnishes more than alters colour, but for both sexes it is evident that the time in the river is taking its toll. At Gavelwood the fish are not as buff as when fresh from the sea. To say they look flabby would be unfair, but clearly huge physiological changes are going on inside them as the fat reserves are depleted to make eggs or milt. Every day they delay spawning the clock is ticking, and just below Pike Pool is a long gravel spit ready for them, scoured and washed by the fast tail of water exiting the pool. Perfect for our much-travelled *salar*, this is Salmon Shallows, where the courtship will take place in full view for anyone who wants to watch.

Courtship is not something that happens every day: if I want to see it I will have to get up to be there at dawn on five to ten consecutive days to see it just once, but in the close confines of a chalkstream it is worth the effort. The weather is no determinant, so don't use that as a judge of when to go. I've seen spawning in everything from lashing rain to snow on the ground. If you had to use a marker the male salmon is your best bet. Close to spawning day he goes neon-red down his flanks and the kype grows so pronounced that he can barely close his mouth. The females on the other hand tarnish just a little more, and like their sister trout, are tasked with digging the redd.

There is no competition with the trout in the shallows; they prefer the relative calm of the North Stream gravel beds to do their spawning. So, soon after sunrise the female salmon drifts back out of Pike Pool to pick her spot a few dozen yards downstream, a flat section of gravel with just enough depth to cover her upright body. It is clever how they pick the optimum spot to dig; if the flow is too fast the indentation soon gets washed flat. Too slow and the female has to do all the digging. Just right and the water flow helps scour the gravel to speed the process.

No sooner has the digging finished than the male appears, in his full tartan finery. In the few inches of water, he really does seem to glow red-hot, arching sideways in the current around the female as she hovers over the redd. Sensing the female is ready, he will take his place beside her, spraying his white milt over her thousands of eggs as they emerge from her body. Caught in the ebb vortex of water the redd creates, the eggs momentarily pause on the current, gently drifting down to settle in the nooks and crannies between the loose gravel. His job now done, the male fins away, leaving the female to kick up the gravel piled around the redd with her tail and body to cover the eggs. Safely trapped in the loose gravel, constantly caressed by the oxygen-rich water of the chalkstream, the eggs will stay safe until spring, when they hatch into alevins, tiny fish with the yolk sac attached, which are the next stage for the new generation of salmon.

The surfeit of eggs, which get carried away on the current or left exposed on the riverbed, brings out the vulture effect in the other river creatures. Winter is a bleak time for them: there is not much food and short daylight hours to find it in, so competition for the eggs is frantic. Fish, eels, crayfish and birds all vie for first pickings, ever alert to the chance of a midwinter protein buzz. My favourite are the crayfish, who display incredible cunning and rare speed. They usually move so slowly as to be barely visible in normal times, but clearly salmon eggs are high on their list of delicacies. Before the salmon even start laying you can see them shuffling into pole position. With their lobster-like claws they are perfectly adapted to picking the sticky eggs from the crevices between the stones, scuttling around until every last one is consumed.

Grayling are the most active fish in the Evitt during the winter because they spawn in early summer, completing their efforts around June. Like the trout and salmon, spawning takes it out of them, so for a few months they lie low in the river pecking order, quietly recovering. But by winter they are fully recovered and with few other fish to compete with become prodigious eaters of both trout and salmon eggs. Shoal fish, they descend as a pack on a freshly covered redd, sucking up the stray eggs and dislodging the gravel with their snouts to throw up others. It is probably for this reason alone that grayling used to be persecuted by game fishermen who saw them as a threat to future trout stocks. The small trout, too young to be spawning themselves, will also take their share, as will the ducks and the moorhens. It is a feast not to be passed up; a nutrient hit that might mean the difference between survival and death in the winter ahead.

As November becomes December the activity in North Stream peters away and it is a good time to see whether the restoration has done its job. The fast flow and clean gravel are all very fine, but without spawning it would be little more than a pretty river. It is redds I want to see, and am therefore delighted that on a bright winter day the sun bounces back off a riverbed pockmarked with them. Impossible to see from the bank, but somewhere beneath the riverbed, gently trapped under the gravel, were thousands of tiny eggs, kept alive and thriving by the cool, clear, oxygen-rich water. For now everything was on track, but until spring arrived success was still in the balance.

It is tradition that I take a walk down the river on Christmas morning. By that time pretty well everything has died back across Gavelwood. The trees and hedges have shed their leaves; the grass is shorn; the cattle that left in October tore the very last ounce of forage from the sward. The vegetation along the banks has all but gone. There is just a tangle of sedge grasses, the brown seed heads of the cow parsley and shrivelled loosestrife along the margin between bank and river. With everything so bleak and open it is the only time of year you get the sense of the scale of the river valley: how very far it reaches and how very flat it is, but most of all how the valley is defined by the river itself.

Beside the path along the riverbank, groups of little dark holes in the ground have been exposed where the grass has been battered by the rain. As I trip along a little fuzzy head pokes up out of one of the holes, surveys me with alarm, and then promptly pops down out of sight. It is a water vole sniffing the air from his winter quarters. Water voles are no hibernators, but without cover and food they retreat from the river during the winter, digging elaborate tunnel homes in the rain-softened banks where they gather in groups. Foragers by instinct, they come out during the day to find food, taking it back to the burrow to share.

But it is a hazardous existence; seven out of ten will not make it to spring. In part this is due to the lack of cover. The holes I see so easily on Christmas Day would be hidden under grass at most other times of the year, and that lack of cover makes them easy pickings for buzzards and owls, particularly as the voles have to range further than they would like to for food. And even in the burrows there are threats, though the voles try to anticipate the worst of these by creating emergency exits to be used in the event of a flood or visit from a predatory mink. I do what I can by controlling the mink population and leaving as much cover as we can for the voles, but life on a river is tough in winter.

As I look down into the water there is no piscatorial Christmas party going on. A near-empty river stares back at me. Not empty in terms of water: with an averagely wet winter the levels are fast approaching their peak. All over Gavelwood we have opened the hatches and removed the boards to let the current scour the silt of summer away. No, it is the absence of fish that is striking. By now spawning is over. The trout have vanished. One moment they are there in front of you, bold as brass in the shallows, and the next they are gone. With the exception of the juvenile trout, and the grayling that will peck away at the surface on the occasional sunny day when there is a hatch of midges, all the others retreat to the deepest, slackest parts of the river to become invisible. Spawning takes its toll. Trout lose anything up to a third of their body weight in winter, and now it is just a question of hunkering down out of sight and danger until the spring.

For the salmon things are desperate. Exhausted by spawning and shrunken in size, strangely they revert to almost the same silver colour and elongated shape in which they arrived in the river. Now called kelts, they turn downstream to head for the sea, carried by the current. They have to leave – their bodies demand it. After weeks and months without food they can only begin to feed again once they reach the ocean. For that is the thing about being anadromous; fresh water is all about spawning and preserving the species regardless of self. Salt water is all about feeding and building up the body to do the former.

It is the females that leave first, immediately after spawning. A few of the males hang around, either fighting or waiting for another female to pair with, further depleting their reserves and reducing the chances of survival. This helps explain why the few fish that return for a second, third or fourth time are usually female.

At Gavelwood we don't see many kelts; when they turn to leave they will be gone in a matter of hours, and there are not many salmon that travel upriver beyond us in the first place. But I feel a huge mixture of emotion for the ones I see. There are clearly some that have only a matter of days to live. Their weakened bodies, with their immune system shot to pieces, are unable to fight off a white fungus that envelops their head, and blinded, they swim sightless in the slack water behind a rock or post until the end comes. For these you can only wish that the end comes quickly. Others are clearly exhausted, disorientated, or maybe both, again sitting in the current, waiting. Waiting for what signal I cannot be sure, but I want to urge them to turn for the ocean where salvation from a death by starvation awaits before it is too late. The last group are the survivors, the ones that just might make it, swimming with purpose, using the current to ease their passage towards the sea. To these you simply want to cheer, doff your cap and wish God speed.

5
SCAR BOY

THE EMPTY MEADOWS, bare trees and the chill winds of February are bleak times for most of us at Gavelwood. The river keepers only venture out when they have to, and for nearly all the creatures, hunkering down out of sight is the wisest choice. But not everyone sees it this way. Down in the near-clear waters of the chalkstream, where the temperature barely changes from winter to summer, the nymphs, shrimps, snails and young fish carry on regardless. In other kinds of rivers, swollen and dirty with winter rains, they might be having a hard time, but here in nature's aquarium, they thrive.

The yearling brown trout is positively spoilt for choice. Free to roam the river whilst the cannibalistic mature fish are in post-spawning torpor our twelve-month-old trout has successfully made the transition from egg to alevin to parr, and now to what is recognizable as a small but perfectly formed brown trout about the size of my index finger. This is the first time I really get to see them, roaming the shallows for food. As alevin, the freshly hatched egg stage where they still have the yolk sac attached to their tiny fish body for food, they stay hidden in the gravel. Once they move to the parr stage they are solitary, staying close to stones and weed for protection. They need to, because only one in twenty makes it through the parr stage of life.

But now, in winter, just past his first birthday, emboldened by his new-found maturity and as yet with few competitors, our little trout carves out his niche in the river, setting a pattern for his life to come. Brown trout are not territorial in the sense of rutting stags that clash antlers; it is more nudge-theory stuff. As a trout you will occupy a space not more than a few feet square that you call your own, which will divide into your feeding spot and resting spot, and it is at that feeding spot on the gravel shallows of North Stream that I go looking for the yearlings today. Pretty well each of the feeding stations in the fast shallow water is taken, with a yearling trout darting side to side for food as it comes by on the current. It is a good place to dine, but dangerous, and there is one particular fish that has caught my eye who is now wise to the dangers, with an upright scar down his flank. A week or so ago when I first spotted him the wound was blood-red, fresh from an attack, but now it is healing. Scar Boy had had a lucky escape, the stab from the heron missing his vulnerable back by a fraction of an inch, with the sharp bill instead sliding down his side slashing the skin open. It must have been a nasty dose of reality to Scar Boy. In the shallows he would feel relatively safe from bigger trout, pike or even otters and mink; no chance of a surprise attack. He knows he is too big for the kingfisher to bother him now, and a cormorant would come off the worse diving into shallow water. No, Scar Boy's greatest danger comes from the heron.

The herons at Gavelwood know me but don't fear me. Swans will stand their ground, kingfishers dart away, but the heron on a slow patrol along the edge of the shallows will halt to coolly watch my advance. They are without a doubt the best fishermen on the river, creeping along or standing motionless for hours on end until the prey comes in range and with a stab of that super-sharp, dirty yellow-orange beak that is covered in black scars, spear the fish. Aside from their stealth, it is the height and colour that give them the edge. At over 3 foot tall they can peer down directly into the water, cutting out the surface glare for perfect vision with a wide arc to sweep across. And that grey colour. You would have thought green was better, but one day I finally got it. Wading along the river, with a steep bank above me, I had this horrible feeling I was being watched. I paused in my casting to look up, and there, almost invisible against the skyline, was a heron, head cocked to one side, watching me. Grey feathers with a white chest. Grey skies with white clouds. It all made perfect sense, and add to that the broken refraction of looking up through water. No fish stands a chance against a determined heron.

For all that, in the winter I always feel a bit sorry for the herons. There are not many, for they are solitary birds, but they just stand there in the rain, their shoulders hunched up like an old man in a dirty grey mac, whilst the water pours off the feathers. As I get close they stretch up on their toes, and with a little jump and sharp flap of the wings, will be airborne. As often as not they will emit a guttural bark – *aarrrchhhh* – as if to express their displeasure, and then glide a hundred yards or so before landing again. They have no natural predators to speak of, and I am certainly not one, but they are cautious enough to keep their distance.

They also know me well enough to know my route is along the riverbank, so they glide inland. This makes me feel a little bit better, because, for a few hours at least, *Ardea cinerea* will be hunting for something other than fish in the shallow ditches that are starting to drown the water meadows. Frogs, snails, mice, eels, insects, worms, rats and voles are just some of the potential items on the menu. In broad terms if it moves, a heron will eat it. They are not great scavengers, but as they weigh in at around just four pounds (about the same as a pheasant) they actually don't need that much food to sustain them through the winter. However, this is of scant comfort to my juvenile trout that represents the perfect meal to sustain any self-respecting heron for a good few days, so when I am gone they soon return to the river.

For Scar Boy life is in part about making uncomfortable choices. He needs to survive but he also needs to feed. Survival is easy enough if he spends all his time in the dark recesses under the bank, but during these weeks and months he has to feed to grow. And to feed he needs to venture out where danger is a daily part of life. I know Scar Boy well enough to instantly pick him out at his feeding station. He has found a big flint that pokes out from the gravel bed. About the size of a house brick, it parts the flow of the water, creating slack water just ahead of the stone. Into this pocket slides Scar Boy, facing upstream ready for food to come his way. His position has become so regular that the gravel beneath where he lies is bright golden, regularly brushed clean of silt by his flicking fins and the deflected current.

Watching him view the food options that come his way, there is no doubt in my mind that Scar Boy has some very certain preferences of what to eat: things he will ignore, things he will go after. Trout eggs are a favourite. Yes, he has no qualms about eating a potential brother or sister. That said, by now the few eggs that become dislodged from the gravel are probably malformed or infertile. So if I see Scar Boy suddenly scoot to grab something on or close to the bottom I'll know it is an egg. The protein hit is worth the effort.

Picking up nymphs, on the other hand, is a far more languid affair, simply because they are so much more prolific. Most days Scar Boy can pick and choose, and so he does. Today it will be mayfly nymphs. Tomorrow bloodworms. Sometimes caddis larvae. On any given day dozens of different types of underwater creatures, at various stages of their lives, will drift his way. Some days Scar Boy will get a fancy for one particular type. Other days he will be a total magpie, going for whatever takes his eye, switching between the different types with no particular pattern. Over the next three months he will chomp his way through over ten thousand water creatures.

That is a lot of tiny creatures to eat, none of which taken individually has much food value. However, Scar Boy succeeds in getting bigger and fatter because he goes on this voracious feeding rampage with a minimum of effort. I can see him sitting in front of his flint eyeing the current ahead, letting the flow gently sway him from side to side. Being a trout with eyes that look directly ahead, he has to use his body to line up with the food, and the moment something comes in sight that he fancies the posture of his body changes. At rest a trout is fluid, his body quite literally going with the flow, gently swaying with the current, but the moment he lines up on prey his body stiffens along its entire length.

Adjusting his height and trajectory with his pectoral fins, the pair on the underside at the front of his belly, Scar Boy lines himself up with the food as it comes his way. Tiny fin adjustments are constantly made, calibrating the speed of the flow, the position of Scar Boy and the path of the tumbling nymph. In slacker water, even if it appreciated the danger it was in, the nymph might be able to swim away, but in the fast water its tiny body is at the mercy of the current. Scar Boy keeps his nose pointed towards the nymph, and when the two are almost touching he opens his mouth to let the river deliver the nymph. Closing his jaws he lets the current in turn carry him back into the pocket of slack water in front of the flint. The deed done, he settles down to await the next victim.

This routine will repeat dozens of times an hour, hundreds a day, but there are days when Scar Boy alters the routine, taking advantage of a winter treat when the sun shines to bring a little warmth to the winter valley, which in turn brings on a hatch of midges. These are not the nasty biting kind; chalkstream midges are equally prolific but more benign. Ever present in the river as first bloodworms and then pupae, which is the stage just prior to hatching in the same way that a chrysalis is to a butterfly, they are at all stages of a life a staple for trout. When they hatch into flies, which they do in January to December, at the slightest uptick in the temperature, clouds of them will gather over the surface of the water.

For Scar Boy the unusual February activity will catch his eye and he will reposition himself to look up at the surface. The thousands of insects are not really his focus – on the wing and flying around they are too tiny to be worth leaping for. No, for Scar Boy it is the still-born ones, trapped in the surface film, that he is after. So for all the time the hatch goes on, as the deaths occur, Scar Boy will hover beneath the surface, sucking down the corpses until a chill descends on the valley, the hatch ends and the snacking treat over, Scar Boy will go back to his lair beneath.

In nature nothing comes by right. The territorial niche Scar Boy carves out as his own inevitably attracts competitors, some trout, some not. The pitch he calls his own does not have to be huge, about the area of a child's paddling pool, because food is so plentiful in a chalkstream. But he guards it as best he can. As a juvenile trout you don't have much in your armoury other than guile, tenacity and the right of occupation to ward off interlopers. Rarely an hour will go by without some sort of ingression into Scar Boy's space, which is either a larger trout, ones about his size or grayling. It is the latter that seem to annoy him most.

Grayling are pack fish, patrolling long sections of the stream in shoals of anywhere from half a dozen to twenty. The pack will include all sizes and generations of grayling, who will surround Scar Boy in the shallows. Outnumbered, but not threatened, there is little Scar Boy can do but put up with them. Occasionally he will express his displeasure by darting in front of one to steal some food, but on the whole it is something of a Mexican stand-off until the grayling, tired by the fast current and the lack of food to feed so many mouths, move on.

Other trout are more problematic. Bigger trout he can do little about, other than defer to their size and wait for the current to make life hard and move them on. In time the same will happen to Scar Boy. He will become too big for the lie, the effort of holding station in the shallows too much for the food reward, so he will vacate his home in front of the flint for a deeper lie with less current. As of now, February, he has no such problems, but come early summer he will be on the move. However when a trespassing trout of Scar Boy's peer group arrives it is another story entirely.

The interloper will come quietly, sidling up to Scar Boy, usually just to one side and slightly ahead of him. For a while he will be ignored and as often as not will leave almost as soon as he has arrived. At other times he stays, but Scar Boy has a short fuse, and after minutes rather than hours he starts to agitate. Scar Boy will swim up beside the interloper as if to say, 'Excuse me, that seat is taken.' At which the errant trout inclines his head – 'I am terribly sorry' – apologizes and leaves. However the gentlemanly approach does not always pay dividends, so Scar Boy uses his tactical advantage. In his position in the slack water ahead of the stone he is sheltered from the fast current; the intruder staying close to Scar Boy hopes to gain the same advantage. Now trout don't like to get really close together. They like to keep a little distance between themselves, so if Scar Boy gradually inches out from the shelter of the stone the other trout will do the same until gradually he finds himself nudged out of the slack water and into the fast current. Disgusted at being so simply outmanoeuvred, the interloper will spin round and swim away, flicking his tail faster than needs be as if to gesture 'Well, I didn't want that spot anyway.' Scar Boy, despite the constant aggravation, will pretty well spend every daylight hour on the shallows, but come the end of the day or if he gets spooked he will retreat to his resting spot in slack water under the bank undercut just a few feet away. He has chosen his hideaway well to give him a view of the shallows so that he can return as soon as the danger has passed or when it is time to start feeding again. There is no doubt that Scar Boy feels at his safest in his hidey-hole. He will switch off, coming as close to sleep as trout ever do, but in this semi-comatose state he becomes vulnerable to his greatest danger.

Pike are one of the ever-constant inhabitants of Gavelwood. We don't have many, which is a good thing in some respects as they are very partial to trout, but for all their fearsome and predatory reputation they do have an important role to play in the everyday life of the river. I will reckon to see one or two each day, but that is not so hard, as they are creatures of habit, occupying the same spot day after day. Shallow, slack water is their favoured lie, and for some reason the confluence of two streams, like where Katherine's Brook joins the Evitt, is a regular hang-out. In the summer the sun will warm the shallows and the pike will languish in the tepid water doing absolutely nothing. No fish, no lure, no form of temptation will break their reverie, but come the winter the same fish, in the same spot, will tell a different story.

The late summer and autumn was not the season for our pike; whilst other fish were feeding in preparation for winter and going about spawning he was biding his time, still recovering from his own breeding efforts that finished in June. Our pike is not a fish prone to expending unnecessary effort. Not for him the picking of tiny insects from the surface as does our trout, or travelling huge distances like the salmon. For *Esox lucius* life is all about waiting for opportunity to come his way in the form of a fish, or frog, or water vole or tiny duckling unfortunate enough to wander across his field of vision at that very moment when hunger is on his mind.

Having remained static for hours, even days, his position in the slack current effortlessly maintained with little movements of his fins, he or she (the biggest pike are usually female) will quite suddenly arch the body sideways and with a powerful thrust of the tail fin will accelerate to grab the prey in a mouth that is a throwback to prehistoric times. Whenever I see a pike I feel compelled to admire the green mottled body, which is lithe and compact, but inevitably my eyes are drawn to the large flat head, with a jaw that accounts for maybe a quarter of the body length. The lower lip protrudes to give the pike that impression of a constant pout, but the ever-seeing eyes, located above and with a direct sightline to the jaws, stare out, and you can just tell that there is a silent vigil involving the triangulation between potential prey, eyes and jaws.

When the moment comes, and it is only a moment, the attack and capture is over in a matter of seconds. What I like best about pike is that after a successful attack they return to the lie and holding the prey sideways in their jaw, sit in the shallows like a self-satisfied retriever waiting for a pat on the head from its master. Of course, they are doing nothing of the sort. The powerful jaws are crushing to death the poor unfortunate trapped in a frightening array of teeth, like a shark's backward-facing, that deny any chance of escape. Once *Esox* has squeezed the life out of his prey, like the kingfisher he rotates the fish around in his jaws, and once the head is facing down his throat, swallows it whole. And that is him done for the next few days.

There is a healthy debate amongst river keepers as to the relative merits of pike. There is no doubt that they eat trout and plenty of them, but on the other hand they are scavengers picking off the weak, diseased and overpopulated. But in the end we tolerate a few big *Esox* in our rivers because they detest competition from jack pike, the young pike, which one day might grow up to challenge them. When it comes to aggression pike save the worst for their own, ruthlessly ethnically cleansing large tracts of river. One or two big pike are infinitely preferable to many small ones.

River life is all about balance, and the decline of Gavelwood has wickedly skewed the equilibrium away from trout, where the population has been in decline for years. The work on North Stream and the restoration to follow will, in time, redress the balance, but for now I have a huge hole in the population with either very young or ageing trout. Of course we have some like Scar Boy who might make it, but on the whole conditions have simply been too tough for the yearlings to make the progression from adolescence to mature breeding stock. So I have a choice to make: let nature play it out or bring in some fish to fill the void. Stocking is one of those highly emotive subjects, and without a doubt stuffing a river full of overly fat, stew pond-raised, pellet-fed trout is hardly part of the chalkstream ideal. But great forests can grow from a few well-tended sprigs, so I went in search of some native Evitt trout.

Rearing trout is one of those things we tend to think of as terribly modern, but since early Victorian times British fish farmers have been exporting native brown trout around the globe to seed rivers with populations that still thrive today. Australia, New Zealand, Chile, Afghanistan, the Lebanon, India, Nepal; the roster of countries runs to dozens, and one of those very same farms, fed by the pristine springs on the chalkstream headwaters, is where I hope to find the fish I need.

Royston has been rearing trout for longer than I have been alive; he would be the first to admit that it is not a hugely complicated process, but it does take enormous care and dedication. It all starts with the brood stock, that live for eleven months of the year like cosseted horses on a stud farm, their every requirement and need satisfied in a pond replenished by a constant stream of water from the spring that flows in one end and out the other. These are big trout, the cream of their generation, anywhere between six and fifteen years old, that gently circle the pond without a care in the world.

But like their fellows in the river, around autumn time the spawning fever takes over. They change from brown, to red to tartan. The males grow a kype. The females get plumper. The swimming around the pond grows more and more aggressive. Daily the farm workers string a deep net across the pond and gather all the fish into a holding circle. Each fish is gently lifted from the water and experienced hands caress and probe the underbelly. Too soft or too hard. It is a difficult judgement to make, but having handled thousands, maybe millions of fish in his lifetime, Royston hardly has to think if a fish is ready. He just knows. The touch is hardly necessary. He can see it as much as feel it.

Ready means gravid, full of eggs for the female or milt for the male. Holding the female Royston squeezes the underbelly and the bright orange eggs, perfectly round, about the size of a peppercorn and glistening in fluid, pour out in their thousands into a container. The last few ova extracted, the fish is returned to the pond for another year, and now it is the turn of the male. Milking his underbelly just above the tail, Royston makes the white fluid squirt out onto the eggs and with a deft spin of his hand mixes the two in the container as if adding cream to coffee. The trick at this point is to get the eggs as quickly as possible into a place that replicates the gravel spawning beds, with a fast flow of oxygen-rich water that will sustain the eggs until they are close to hatching.

Sometime in the nineteenth century Royston's farm had sunk a borehole deep into the chalk aquifer below. Today it still flows, the constant supply feeding a line of upright pipes that emerge from the floor of an old barn. The cold clear water bubbles out over the top of the pipes, pouring onto the ground and out through a ramshackle door, keeping the place ever wet and damp. Briefly cutting off the supply of water, Royston pours the fertilized eggs into one of the tubes until it is two-thirds full, places a grille over the aperture at the top and turns the water back on. Peering down through the grille I watch the eggs tumble and dance in the flow like balls in a bingo machine, and that is where they stay for about two weeks until they get a little black spot inside the egg. This is the eyed stage, and that spot is the first indication of a tiny trout.

Royston's fish farm is not exactly what you would call state-of-the-art; nobody ever set out to build this for rearing fish. It has always been a farm, but of the more traditional kind, with hay barns, a milking parlour, cattle sheds and Royston's home, the old farmhouse, in the midst of it all. Stretching back along the bank of the river, covering three or four acres, are various man-dug ponds and streams, fed by water from the river, that are all full of trout at different stages of development. If you ever want a God complex visit a trout farm – the fish greet all visitors as food providers. Stand at the edge of the pond and they will leap, twirl and churn the water at your feet until you throw in a scoop of food.

As Royston takes me from one damp building to another I blink to adjust my eyes to the dark of what used to be the milking parlour, converted to house four raceways that run the length of the building. The raceways are constantly fed with water from the same borehole that feeds the upright pipes. Here the newly eyed eggs are transferred to mesh trays which are in turn slotted into the raceways immersed in a constant stream of water. Each day the black eye in the orange egg gets bigger and each day the farm workers scour the trays to check all the eggs are healthy. The ones that die turn white and are gently removed with a pipette. The dark. The water. The constant temperature. It is all about trying to perfectly replicate life in a river.

It is roughly three months on from fertilization when the eggs morph into tiny trout, albeit with the egg sac beneath them. This alevin stage lasts another three weeks until the sac shrivels away to leave a tiny fry about an inch long. Further down the raceways are deeper tanks, into which thousands of these tiny fry are transferred, swimming in groups that look like blots of ink in the water. In the wild they would feed on plankton, but here at the farm the daily feeds of dust, an incredibly fine protein food finer than flour, sustain them until they are ready for the move outside at four months after hatching. Outside in the ponds Royston shows me trout of every age, from a few months to six years. They are not all big; in fact they mostly are not. It takes a brown trout two years to get to 10 inches or half a pound. No, it is not the size of these fish that is the fascination for me, but rather the fact that every single one is a blood brother of the fish I see each day at Gavelwood. The brood stock Royston so lovingly handles were either captured from the Evitt or their forebears were, and with these fish I can fill in the voids left by the years of neglect.

There is no absolute science to stocking a river; it is a matter of judgement, experience and educated guesswork. From what I had seen from the spawning activity we seemed to have plenty of mature, older fish. Scar Boy and his cohorts were in enough numbers to be confident that the juvenile population would thrive with the benefit of the restoration. The voids were in the two-, three- and four-year-old fish. Sitting at Royston's kitchen table we made a stab at the numbers of fish that should be in the river if everything was perfect and worked back from that to arrive at some numbers.

Royston and I agreed on the first Monday in April for the fish to be brought to Gavelwood. I took a last turn around the farm to check on them, this time feeling a little paternal as they again leapt and twirled at the sight of me. But I left still wondering whether this was the right decision. Ultimately only time would tell.

6
MARCH

T. S. ELIOT SAID April was the cruellest month; he clearly knew nothing of life on the river in March. For most of the river creatures there is no easy living. Food is short and body reserves at their lowest ebb. The valley looks desolate; hardly anyone or anything ventures out. Nearly every form of fishing is closed. The trout keep hidden, the grayling move into spawning mode and even the pike are out of season.

Occasionally a salmon angler will tough it out, but you can tell it is a token visit. A day to breathe damp air and get out of the house. The chances of catching an Evitt salmon are so remote as to be infinitesimal. The catch record book goes back over a century. Admittedly the entries are sporadic and in some years non-existent, but nowhere can I see a March salmon recorded. I would say that it is an exercise in hope over expectation, but even that overstates the expectations of these few, solitary anglers who know in their heart of hearts that the arrival of *salar* from the sea is still months away. But they do it because sometimes it is enough to stare at the ever-rolling stream, be a part of it by fishing and dream of better days to come.

But however foul the weather, bleak the outlook or miserable the work in prospect, I treasure March days. These are the last when I will have the place to myself. Once the fishing season starts in April people intrude on the landscape. Is that selfish? Probably. So every dawn morning that breaks bright and clear is mine. The valley stretches out for miles in every direction. Distant church spires usually hidden by trees poke up on the horizon. The gentle rise of the land from the river, across the meadows and on up to the denuded woodland, in most months obscured by the hedgerows, is for once apparent. The course of the river, in summer rough-edged by overflowing vegetation that tumbles from the bank, is now bare. The join between land and water is perfectly delineated, the river looking wide, clean and powerful as it drives through with winter pace.

The time has come for me to make the final push to prepare the river for the season ahead. The jobs I prevaricated over or simply put off for no good reason loom on the calendar. Banks to repair, fences to mend, fringes to cut, fallen trees to clear, potholes in tracks to be filled, machinery to service, sluices and hatches to tweak to get the perfect flow. No single job in itself critical, but as a whole the work that will make Gavelwood a better place for the creatures who watch me go about the daily tasks.

If you were that barn owl that I see taking a dusk or dawn hunting flight across the landscape of the chalkstream valley you would be very aware of how the landscape changes beneath you every few yards, the changes radiating out from the river. Ecologists call this cline. The dictionary definition is 'a continuum with an infinite number of gradations from one extreme to the other', which translated into the life of Gavelwood describes the change from the extreme wetness of river margins to the dry woodland well above the water table. As that owl you care little about how wet or dry the ground may be; after all you barely touch it when you swoop on your prey. Nor will you care very much about the vegetation that grows from the ground, be it tall green stuff in the wet or the stubblier, shorter growth in the dry. All you will know is that with each gradation there is a different type of small-scale habitat that holds a different meal from furry field mouse to slimy toad, with everything in between. It is this cline, created first by nature but now cosseted by man, that makes the chalkstream valley so extraordinarily diverse.

I start my March list of jobs on the riverbank cutting down what is generally referred to by river keepers as the 'fringe', which is the swathe of growth along the bank that separates the river from a mown path. In summer this is the most amazing collection of native plants, with names that are fantastical: reed sweetgrass, comfrey, fleabane, marsh woundwort, ragged robin, hemp agrimony (Latin name *Eupatorium cannabinum*, so no prizes for guessing its origins) or hemlock water dropwort to name just a few of the common ones. They are not dull plants to look at either. Each meadowsweet has a host of fluffy flowers that look like lamb tails. The water avens is reminiscent of the purple fritillaries that are lovingly nurtured in the 'wild' sections of formal gardens. Comfrey, with elongated bell-shaped flowers in white and purple, is a magnet for bumblebees.

The river fringe is, believe it or not, one of the hotly debated aspects of river management. Every keeper harbours an opinion. Some like it narrow, maybe a foot wide and trimmed short to knee height. Others like it wild and wide (up to 2 yards), with just the worst of the excesses trimmed off. Which is right? Well, at either extreme neither does the job properly. Cut back to nearly nothing it deprives insects and wildlife of a home along the river's edge. Left to run wild, the dominant, invasive plants crowd out the others and fishermen cannot reach the water. The one thing everyone agrees on is that March is the time to cut back the dead growth from the previous year, but even then it is no scorched-earth regime. The riverbank, with all its chaotic growth that tumbles over into the water, is the vital connection between the river and the valley through which it flows.

For the creatures it is many things; a home for a nest, a profitable hunting ground, a handy staging post when moving from the river into the meadows beyond.

Cutting, even on a cold March day, soon works up a sweat; it is surprising how much force you have to exert to swing the brush cutter through the tough stems of the tall, spindly dead plants and matted grasses. As I took a break sitting on a stile I saw the layer of newly cut stems start to move and suddenly this little face popped up from beneath. The ears, disproportionately large for the head, swivelled in my direction, assessed me for what I was (no threat) and disappeared beneath the cuttings. Clearly on a mission, I could trace the stop–start zigzag progress of the field mouse as he made his way along the bank by the movement in the cuttings. My work had created an unexpected bounty for *Apodemus sylvaticus* – wood mouse, or long-tailed field mouse – at an important time of year, as the mice are just getting into the breeding season. The bounty is all the seeds dislodged by the cutting, and the mice were soon out in force, scurrying around to eat and gather as much as they could.

The field mice are funny little things, I'd hazard the commonest mammal at Gavelwood – they are nearly everywhere. Along the banks, all over the meadows, in the hedgerows – about the only place you won't find them much is deep in the woods. Every day I will see them scampering or hear the rustle of the dried grass as they head for cover. The food they eat is an everyday staple of the valley – seeds, fungi, worms, tiny insects and even snails. I am no threat to them and today I am their saviour. Despite the noise, commotion and fuel fumes from the cutter they almost get under my feet in their eagerness for this unexpected feast. Once they have eaten their fill they start to carry off twigs to build the nest for the first litter of the season that will shortly be born. It's a pity that not all of them will make it. Though I see plenty of field mice during the day, they are mostly nocturnal creatures. The huge black eyes are adapted to suck in as much light as possible, the ears attuned to every sound, and the pointed nose has a sharp sense of smell for locating food. So far so good, but the one device they lack is radar, and that is a shame, as their greatest enemy comes from the sky in the form of my dawn hunting companion the barn owl, whose night-sight trumps that of the field mouse.

Barn owls just drift across the valley. There is no hurry to their lives. Bats are frenetic, raptors tense and ready, but owls take it all very steady, wings stretched out on the air with head down scouring the ground below. If they make the stoop for a successful kill, they head off, the victim dangling from the beak, with no sort of triumph, to share the spoils back at the nest. If they miss out they'll just perch on a nearby fence post or tree branch for a while, then resume the hunt. I have no idea why barn owls are so languid. Maybe it is because with so much food on offer most times of the year (water voles, baby rabbits and shrews are always on the menu) they have few worries about starving. Tonight they certainly won't. I know for sure that at dusk the newly cut bank, now a magnet for the field mice, will become a corridor of plenty for the owls. That is the thing about managing a river; one action brings many reactions.

Sometimes I don't spend enough time taking a really good look at the entire length of Gavelwood, but a few days' bank cutting seems to be nature's way of imposing some discipline on me to gain a better idea of how every yard has coped with the winter. Chalkstreams are not like big spate rivers where exceptional floods can move boulders the size of a house or cut a completely new path for the river. The changes are more subtle and gradual. A fallen tree may have redirected the current. A bank collapsed after years of unseen erosion. If we have had a dry winter the riverbed will have silted up in places for lack of a flush.

Yard by yard I take it all in. There is not much insect life on the surface to distract me; an occasional hatch of midges, but it is fleeting. Sometimes if the sun shines for a while around the middle of the day there will be a hatch of large dark olives. The fly-fisherman in me will instantly perk up at the sight of these up-winged flies, with their brown camo body and translucent veined wings that catch the sun's rays and positively shine as they bob on the surface. These are the type of flies that trout like to eat, not the biggest a trout will see in a year, but ten times bigger than the midge option. If the trout start to rise I curse the fact that the opening day is still some weeks away, though in truth with the adult trout pretty well dormant the olives will usually go unmolested unless Scar Boy and his fellows in the shallows get a taste for them. The real March action is under the surface.

Assuming that the water in the river remains at a fairly constant temperature all year round, which it does, it is a fair question to ask by what calendar the insects measure their year. What is to stop a caddis nymph, a fly that usually hatches as sedge in July, mistaking a hot April as the right time? Why is it that every fishing record book going back centuries notes the start of the mayfly hatch on the chalkstreams in the second or third week in May? Across those centuries the months leading up to May have been wickedly cold, exceptionally snowy, the driest since records began – you name it, those extremes have happened – but still the hatch comes bang on time. It is tempting to think that it might be a particular weather window that triggers the hatch, but *Ephemera danica* emerges just the same on a cold and squally day as on a hot and sunny one.

The explanation lies in the length of the day, the time from sunrise to sunset and the associated light intensity. This is the one constant for the past few million years, and the chalkstream insects have been there pretty well right from the beginning, or at least 318 million years according to fossilized mayfly remains. So whatever the weather, the sun has always risen at the same time and set at the same time, and this is the calendar to which the insects attune their bodies. March might seem a mostly miserable month for anyone on the riverbank; for the aquatic creatures they are already closer to the longest day than the shortest, where under the water they are evolving in a steady progression ready for their big day. That day can be anything from a few months to two years from the moment that the spinner, the egg-laying female, dips down to deposit her eggs on the surface of the river. For the ubiquitous midges that I see hatching every month of the year, the cycle from egg to nymph to midge is two months. For the mayfly the same cycle is two years, and when my March companion the large dark olive finally breaks surface it is the end of a six-month evolution.

Try as I might I have never seen olive eggs, or any other insect egg for that matter, in a river – they are so tiny as to only be observable through a microscope. It seems incredible that something so small can withstand the drift downriver until the sticky eggs randomly attach themselves to the gravel, weed or river plants. But they do survive this tumbling lottery, and in great numbers. There are a few concerned parents like the *Baetis* species who take the trouble to crawl down into the water using aquatic plants as a ladder so that they can lay the eggs directly on something in the river, but these are most definitely the exception rather than the rule. Our dark olive is one of these fortunate offspring, but as soon as he hatches from egg to nymph, like an ungrateful teenager he will abandon his birthplace and spend his days darting from stone to stone or flitting amongst the cover of weed beds. As he grows he constantly moults, bursting out of one old skin after another, getting larger every time. Food is easy to find. The stones and weed have a slimy covering of algae, a community of tiny plants that the nymph can graze at will. Not all his fellow nymphs are quite as peripatetic. Some, like the mayfly nymph, tunnel into the silt. Others fall into the categories of silt crawlers or moss creepers, which fairly well sums them up, or there is a select group who spend their entire nymphal life clinging to a stone until the day they launch themselves towards the surface.

Whatever the habits of a particular nymph it is around this time of year that their abundance in the river starts to soar almost exponentially as they moult and mature ahead of the prime hatching months of April to July. For the chalkstream angler there is definitely a sequence to tick off as the season progresses: the large olive in March, grannom in April, mayfly in May, pale watery spinner in June, blue-winged olive in July, and so on. But this list does no real justice to the huge diversity of insects in the river, probably running to thousands, which are not only an entomologist's delight but a feast for the fish for which nymphs and the other underwater creatures make up nine-tenths of everything they eat. For the angler who prefers the dry fly, imitating the hatches of insects alighting on the surface, this is a salutary statistic.

Trout and all the other fish in the river like nymphs for two reasons: there are masses of them and they are easy to eat. For us standing on the bank looking into the river, the vast population is hard to see, but run a fine-mesh net through the water or turn over a stone and a whole new world is revealed. Nymphs are everywhere, and if they were larger, say even the size of a mouse, you would throw the net down and run screaming. Up close they could be prehistoric monsters, given that they have barely evolved in hundreds of millions of years. They have a segmented body and six articulated legs, overall a bit reminiscent of a scorpion, but without the curved tail. The head is bulbous, with large wide-set eyes like a housefly, with a powerful mandible. At the back, three tails give it a total length of anywhere from an eighth of an inch to over an inch. The crawlers, creepers, clingers and diggers have strong legs, but the swimmers don't, propelling themselves through the water by wriggling up and down like a person executing the butterfly stroke.

For the trout it is these swimming nymphs that they prefer to feed on, and once you know the pattern it is a fascinating little cameo performance. Nymphs are too small and not enough of a nutrient hit to be worth chasing after, but with so many to feed on trout would be foolish to ignore them. So look for a trout sitting horizontal in the water, facing directly upstream. Every so often he drifts to the left or the right, opens his mouth, swallows and returns to his holding spot. That is a trout feeding on nymphs. The key to the observation is the mouth, which shows out bright white from the inside when opened. This moment is what the famous river keeper Frank Sawyer, also the inventor of the most successful fly of all the time, the Pheasant Tail (a nymph imitation), described as the 'chink of light'. If you are fishing, that moment of the 'chink' is the time to raise your rod tip and strike before the trout works out that your imitation on a hook is not the real thing and spits it out. The clingers, burrowers and crawlers may mostly escape the attention of the trout, but there are plenty of other open mouths ready to take these on.

The term 'biomass' is regularly bandied about by environmental scientists; it is an ugly phrase used to describe the total weight of all the fish in the river, from largest to smallest; at Gavelwood, among the regular inhabitants that would go from pike down to sticklebacks, plus everything in between. Counter-intuitively, it is one of the smallest of the fishes, the bullhead, that accounts for a quarter of the fish in the river by weight and an even higher proportion by numbers – only the sticklebacks are more populous. Faced with these sort of numbers arrayed against them, the clinging and crawling nymphs have a tough battle for survival.

For a fish so common, you'd think that the bullhead would be easy to spot: after all it is hardly what you would call microscopic, growing as it does to 5 inches, though more commonly 3. It is also pretty distinctive, unlike any other river dweller, with a blackened, flattened head, reminiscent of a bruised miller's digit after it has been caught in the grinding mechanism (ouch!), giving rise to its other name, the miller's thumb. The truth is that of all the fish in a chalkstream it is the one with the best camouflage and the most inconspicuous lifestyle, living as it does under stones, only coming out to feed at night. Perfectly adapted to life in a fast-flowing stream, its large pectoral fins create downforce to hold it steady on the bottom; the nymphs under and around the gravel bottom are easy pickings, but really everything from tiny fish to their own eggs is fair game for which they compete fiercely with their fellow bullheads. They have a terrible reputation with river keepers as devourers of trout eggs, but they are as much sinned against as sinners. At certain times of the year trout will eat nothing but bullheads, providing 80 per cent of their protein needs. Kingfishers devour them for a pastime, as do pike and herons, who are clearly all ignorant of the international conservation status of the bullhead that puts it on the same level as the Atlantic salmon.

All in all, life in the chalkstream food chain is not without risk, which might explain why the smallest of them all, the stickleback, is the most aggressive of them all.

The three-spined stickleback (*Gasterosteus aculeatus*) are the fish of our childhood, the ones we netted from ponds and ditches, displaying them proudly to our parents in jars as they darted hither and thither clearly angered at their sudden relocation and confinement. For a stickleback does anger, along with vanity and courtship, quite unlike any other freshwater fish. For a start it is armour-plated, far tougher than any river fish needs to be. Its skin, and for that matter its general appearance, is closer to a tuna than a trout. Along its back, instead of a dorsal fin, there are three wickedly sharp spines, from which it get its name. Just above the head there are three more spines, two facing forwards and one back, and to complete the top, front and bottom array of weaponry it has a final short, sharp spine under the belly. By early summer the warm back eddies of Gavelwood will be thick with schools of inch-long sticklebacks, flitting together in a swirling ball. But wind back to March and it is a very different story.

The mating ritual of both trout and salmon, however selective choosing the mate may be, looks a cursory affair, with little courtship or sense of occasion – some might say it is cold-blooded. However the stickleback is very much the Casanova matador, dressed up in finery that sets him apart from every other fish in the river. He is as bright and dapper as the kingfisher, and the more the mating ritual progresses the more his fishy plumage glows. The start is to carve out his territory; there is no great science to this. He picks his square foot of riverbed and then protects it jealously from all comers. When it comes to territorial protection this is not the subtle style of war. Our male believes in the full-frontal approach, swimming fast and direct at the invader. If they do not retreat he stabs at them with his spines or bites them. The more he wins the battles, the brighter grows the hue of his silver/green flanks. The pale pink blush under his throat goes a bright red. Badly injured invaders slink off, visibly turning grey in defeat.

The territory secured, it is time to build the love nest. Using his spines as forks, the victor rolls around in the riverbed to create an indentation that he now lines with short lengths of reed or twigs that he binds together by secreting a slimy mucus over them. The floor finished, he adds sides, one end and a roof. The nest complete, a tunnel an inch in diameter and about 3 inches long, it is time to find a mate. This involves lots of swimming around in front of prospective wives; presumably his super-bright appearance indicates that he is a valiant fighter and successful nest builder, so soon a female will peel off from the group and the male will usher her towards the nest. Once inside she will lay a few eggs, indicating her success by breaking through the end wall and swimming away. The male will follow into the nest, cover the eggs with milt and then stand guard until the following morning, maintaining his routine of feeding and attacking all comers. The next morning the whole courtship process resumes, culminating with more egg-laying and milting. Whether it is the same female every time who can tell, but after a few days, when the male decides he has sufficiently filled the nest he boards up either end.

You might think that this marks the end of the stickleback's paternity interest – after all every other fish in the river abandons the eggs, and usually without this level of protection. But not the stickleback. He stands guard for anywhere between two to three weeks whilst the eggs hatch. Although it might be tempting to think that standing guard is superfluous, should he meet an untimely end other sticklebacks will shred the nest and devour the contents. But assuming all goes well, dad will keep the newly hatched sticklebacks confined to the nest, chewing up food for them to eat, and at around a week old he will dismantle the nest, the family living in the ruins for another week under his guardianship until they gradually start to disperse.

Over the years, the more I see the sticklebacks the more I have come to admire their sheer determination to perpetuate the species. I guess being so small you have to try that much harder, and stay small they most certainly do. Most fish keep growing if the food and habitat allow it: pike to thirty or forty pounds, trout to six to ten, grayling to three. But the little stickleback remains in that fraction-of-an-ounce territory however abundant the food or good life gets. It could just be that nature recognized that with his warring lifestyle and pumped-up body, if he got any larger he would wipe out every other fish in the river. Yes, the stickleback is best kept small.

By the end of March there is just a hint of spring about Gavelwood; a very slight hue of green to the trees and hedges. Some of the woody plants, like nettles and hogweed, are pushing up shoots along the riverbank where the relatively warm water keeps the frost at bay. But it is the water in the river itself that is my greatest concern.

Such is the nature of a chalkstream that for good or ill, by this moment of year the die is cast for the next six months ahead. Great, bad or indifferent, with one look at the river I can predict the season to come, and it is all down to the unique geology that feeds water to the chalk rivers.

Chalkstreams are not confined to southern England as is often supposed. Yes, the most famous and the majority of them are crowded into Hampshire, Wiltshire and Dorset, but they flow in the continuation of chalk strata that starts north of Hull in East Yorkshire. The seam heads down the east coast, goes briefly into Norfolk and then turns southwest, skirting London across the Chilterns and then on towards Southampton and the Jurassic Coast, where it dives beneath the Channel to re-emerge in Normandy. Here the Risle, the Charentonne and the Andelle are the epicentre of French fly-fishing, the traditions, styles and hatches of England reproduced with a Gallic twist. Charles Ritz, he of hotel fame and mentor to Frank Sawyer, made them famous in the post-Second World War era, and it's the only place I've heard of Hemingway fishing a chalkstream.

It seems an obvious thing to say, but chalkstreams are unique and special not because they flow through chalk land, but because the water in them has come from a natural chalk reservoir deep beneath the ground. Put at its very simplest, the water I watch flow by in the Evitt today fell as rain months ago, on downland many miles away. That rain filtered into the earth, eventually finding the chalk seam that is thousands of feet below ground. That seam then heads towards the sea, and as it does it gets closer to the surface until the water breaks out into millions and billions of springs that gather to create the stream. The geology of the process is, inevitably, more complicated than that. In some river catchments, the journey from raindrop to river water lasts just a few miles and a few weeks, but for most it is months or years. So think of the chalk as a giant sponge and you are some way to understanding how the process works. Take that dry sponge and gradually pour water into it. For a while the sponge will keep absorbing the water until at some point you have saturated it. From now onwards, for every new drop you pour in an equal amount will flow out. That's how it works.

I am no soothsayer; my confident prediction for the season ahead is based solely on the rain that has fallen over the winter. There is an old saying in river-keeper circles that the only good rain is the rain that falls before St Valentine's Day. This crude but effective rule of thumb is underlining the truth that the rain that falls in, say, May will only reach the river in October, by which time the season is over. The future is governed by the past – one drop in, one drop out.

Of course one dry winter is not a disaster. If that was the case the chalkstreams would have vanished long ago. There is enough tolerance in the geology to cope with the regular variations of the English climate, and sometimes after such a winter you would be hard pushed to see any change in the river over one summer. The water will be as pellucid as ever, the insect hatches follow their regular timetable and the trout go about their daily business. The real change will be to the velocity of the water as it pushes downriver, which will gradually slow without the aquifers pumping away at their normal rate. Think of it as turning the tap on your garden sprinkler down half a turn; appreciable but not critical. That is the effect of a single dry winter. But turn it down another half-turn, then another, and keep up the half-turns until the point at which the sprinkler stops working, then you would have the effect of successive dry winters or abstraction by man – a river that gradually stops working.

But chalkstreams are amazing things; freaks of nature that have survived thousands of years with water so pure and so clear that put in a glass it is as good (or better?!) as anything you can buy in a bottle. Every day millions of gallons flow up from the aquifers, down the rivers and into the ocean, and in that fleeting journey create a home for a unique collection of creatures that simply would not exist as they do without that water. However, all rivers are fed by rain or something similar like snow melt, so what makes a chalkstream so different? In the end it is that slow release mechanism of the chalk sponge that makes the difference, filtering, changing and chilling the water. That filtering effect through the tiny fissures in the chalk is critical for creating the gin-clear water, free from any soil or particles in suspension.

As the water touches the chalk its pH changes from the slightly acidic quality of rain to the alkaline of the finished chalkstream water, and after its time deep underground the water emerges at a cool temperature just perfect for trout and the invertebrates. Over the years, the more I have come to understand how chalkstreams exist the more amazing and special they become, but they can only exist today because a long, long time ago two events occurred, entirely unconnected and 50 million years apart.

The first event was a slow burn, lasting 20 million years during the wonderfully named Cretaceous period. England was beneath the sea for all this time, and as layer after layer of dead seashells and tiny marine plants were compressed and crushed, the chalk seam was created. As the water retreated the land emerged, but what it left behind was nothing special. It took the second event, the shock waves from movement of the tectonic plates that forced the Alps into being 15 million years ago, to remodel England with those rolling hills, the downs, that now capture the raindrops. Even after all this geological upheaval there were no rivers, chalkstream or otherwise. It took the ice age that ended around 9000 BC to put the finishing touches to the landscape, but we are still some way from the creation of the River Evitt and its like.

After the ice age everything became incredibly wet and covered in woodland, a sort of English rainforest if you like. But still there were no rivers; it was just a swampy landscape into which came a creature, extinct in Britain today, whose dominance would result in creating the rivers. That creature was the beaver, a mammal whose obsessive dam-building gradually turned rivulets into rivers. As the swamp retreated the Iron Age people began to clear the woodland to cultivate the fertile valley soil, and around the time of Christ the first water-powered grinding mechanism came into being. To call it a watermill would overstate its influence on the river landscape. That change was left to the Romans who imported their considerable expertise to harness the power of water, and it is probably due to their legacy that a dozen or more mills are recorded in the Domesday Book for the River Evitt alone – there would have been thousands on the chalkstreams all told, and from this point onwards the valleys began to look something like they look today.

And as for today, the last of March and the day before the fishing season starts, I have every reason to be hopeful for the spring and summer ahead. The winter was wet, the work is done, and as the landscape sheds its winter veil I take a last turn around Gavelwood to check that all is well. Tomorrow I return as a fisherman.

7

HOW I HELD A TROUT FOR WARMTH

I CAN'T EXPLAIN why, but I have this tradition that I have to catch a fish on the opening day of the season. To just go fishing is not enough; a fish must be caught, and only a trout will do. Yesterday, the first of April was that day, and by the end of it I can truly say that never before had I held a fish in my hand and thought, 'Goodness, that feels warm.'

It would be a pleasure to report that the opening day at Gavelwood was one of those rare bright sunny early April days, with just a light breeze and a scatter of clouds. The sort of day where you can shed your outer clothing and enjoy the spring sun warming your back, whilst a regular hatch of insects brings the river alive for a few hours either side of noon. Heaven. Sadly not. It was a wicked day, with a cold north wind whipping down the valley registering a chill factor of –4C. It was a ridiculous notion to go fishing at all. That old adage 'When the wind is from the north only the foolish angler sallies forth' rang in my ears as I pulled on thermals and so many layers that it was like casting while wearing a plaster cast.

Naturally the opening day's fish has to be on a dry fly. To the non-fisher this might seem a ridiculous notion – what does it matter on what you catch the fish, just so long as you catch? But matter it does. Once you have a notion in your head, however illogical, doing anything else seems as pointless as cheating at patience. So I dutifully tied on a Large Dark Olive; there were no olives around to imitate, but for that time of year it is the most likely insect you will see fish feeding on.

I could pretend that I waited for a fish to rise, cast the fly that landed like thistledown, drifted serenely on the surface and bang, there was my first trout of the year, but that was not to be. I could say I logically cast left, right and middle in turn, covering every inch of water, but the weather put paid to that as I slogged away into the teeth of the wind up the main river. After half a mile I blamed the fly and changed to a Parachute Adams, one of my 'go to' dry flies that rarely fails, that looks like all sorts of flies fish like to eat. Not that I didn't see plenty of fish as I waded up, but they were all ones that fled as I spooked them from the cover of the river weed. Things were getting desperate as I reached the top of the river, with my score sheet still blank.

In the lee of the Drowners House I considered my options as the prospect of being skunked, going home fishless, looked a likely but shameful outcome. I was really too cold to be past caring – the water and wind soon suck the heat out of your body – but pride is a wicked taskmaster. I am sure there have been opening days in the past where the fish have won out, though today, unaccountably (maybe it was the cold?), I could not recall any. So as a little feeling came back into my hands and legs I decided to give it one last shot on the North Stream.

The current in the North Stream is appreciably slower than the main river at this time of year, so with less depth and turbulence the fish are easier to spot, and sure enough there were a few good-size ones sitting just off the tail of the shallow gravel runs. I fished at each one I saw in turn, but none showed a flicker of interest in the fly bobbing on the surface, way above their heads. So on the principle of if the mountain will not come to Mohammed, it was time to abandon my purist principles and fish a nymph that would sink down to where the fish were feeding close to the riverbed. With my first cast, before the Pheasant Tail Nymph even had a chance to sink very deep, a tiny fish grabbed for it. Mercifully in my frozen state I failed to react, and he vanished before I could hook him. A few yards further up I spied a much bigger fish, definitely worthy of being the first of the year. Still lithe from a sparse winter, with not an ounce of extra fat, his red spots glowed against his pale silver-brown skin as he let the current gently sway his body with the flow. He was definitely ready to feed, so I started to work on him, drifting the nymph close to his lie.

A few times he finned across to take a look at my offering, almost poking at it with his nose, but his mouth he kept resolutely closed. After a few passes I was about to give up, but as a last resort I tried a trick as old as fly-fishing itself – the induced take. That is to say, as the nymph enters the trout's field of vision you move it in the water to imitate the prey fleeing from danger. It makes no sense really; after all the trout had already ignored the same fly half a dozen times. But work it does. Maybe the sudden movement fires up some latent aggression or perhaps it trumps the natural caution of the trout to weigh up all the options, but whatever the reason the fish turned back to chase the fly. He grabbed it with such force I barely needed to strike, and from that moment my opening day duck was broken. A minute later as I clutched him in my frozen fingers to remove the hook he felt warm to the touch, a positive hand warmer. As he swam off I'm sure he was as glad to be away from my cold embrace as I was to be heading home.

Around this time of year it is the nesting birds of Gavelwood that are the most active of all the residents on the river itself. The largest are the swans; we just have a single pair that has been here for years. Next in line are the geese, but really they are more occasional visitors, coming then going but always at war with the swans. Mallards are the resident ducks, and at the bottom of the pile, constantly beleaguered and put upon by all the others, are the moorhens.

Try as I might I can't feel any great affection for the swans, even though they are a fixture and know me every bit as well as I know them. It is ridiculous to say it out loud (they are after all just birds), but they don't seem to have any humanity. During the nesting season they bristle with aggression and for the rest of the year they are simply defiant, pushing up and down the river defending their patch. I should be grateful of course; without the nesting pair our river would become a refuge for dozens and dozens of displaced adolescent swans for the entire summer as they fight, defecate and generally create mayhem before flying off to find a mate.

It is often said that swans mate for life. I don't know if it is absolutely true in every case, but it does seem that way with the pairs I see. Of the creatures at Gavelwood it makes them pretty well unique; fish have no such loyalty. The female mayfly will struggle to stay aloft gripped by two males, and as for the mallards, well it is on the wrong side of necessary, with as many as six drakes mounting one hen in rapid succession. I am not sure the hen always survives, such is the violence of the pursuit and consummation, but the result is huge clutches of eggs, anywhere up to fourteen.

It is tempting to think that the mallard lays so many eggs on the basis that only a few will hatch, and sure enough during April, May and June the riverbank will be strewn with the empty blue shells, stolen and licked clean by stoats. But assuming the eggs don't get eaten it is very possible for every one to hatch, and I regularly see the mother leading a brood of a dozen or more day-old chicks up and down the river, who are able to swim from the moment of their birth. However, it is at this point the real attrition sets in.

I can almost tick the brood size down a chick and a day at a time. Twelve on day one, eleven on day two and so on, until after about a week it will stabilize at around three or four once the predators have taken their toll. The chicks are assailed from above and below. From high above the barn owl will swoop down if the brood crosses open ground. A heron hidden in the reeds will pluck one from the gaggle swimming by, and the pike is never averse to grabbing one from below. Occasionally the mother will sense the danger and flip into a sacrificial defence strategy I only see in ducks. She'll abandon the chicks, swimming away across the river surface clumsily beating her wings against the water, creating as much commotion as she can. It seems to me she is pretending to be injured, saying to the predator, 'Take me, I'm easy pickings', and all the while drawing attention away from the chicks.

The mother always tries her best to protect the chicks, herding them into a tight group amongst the reeds if danger threatens, but it is pretty clear a big brood will rarely stay big. As much as anything else there is always the chick that gets separated. In the world of the mallard 10 yards is as good as 10 miles as the plaintive squeaks of the lost chick are ignored by the mother who concentrates on the needs of the many rather than the few. Sometimes the little chick will get his bearings and summon enough strength to make it back to the brood, the relief in his little body palpable as he swims into the safety of his siblings. But more often than not he will pointlessly swim hither and thither, racked with panic, checking out first the reeds then the open water. But it will be a fruitless search for his brothers and sisters that gradually depletes his strength. Without the warmth and guidance of his mother he will be dead the following morning. The swans on the other hand have no such survival problems; if they have five cygnets it is a fair bet that all five will make it through the end of the summer.

Swans are not only loyal but they are birds of habit, returning to the same nest year after year, which at Gavelwood is in a thick section of reeds just below Pike Corner. Most months of the year you would walk past it and not give it much thought: a large mound of desiccated reeds with a slight indentation in the middle. But come April you'll know it is there before you see it, as the swans will hiss with aggressive intent at your approach.

Being the good partners that they are, swans build the nest together. The cob, the male, tears up reeds that he passes to the pen, who layers them in a circle. Over the years it gets taller and taller as they add to the nest, the annual gradations discernible by the darkening of the reeds like the rings of a tree. Swans choose their material well; this is the same reed that was used by man for thatch in ancient times. Our pair starts the rebuild in March, which is ahead of other swans across the country. They can do this because living as they do at water level on a chalkstream they are protected from the worst of the cold, as the relative heat of the water creates a warm greenhouse layer from which they rarely stray.

By the time I venture out to break my fishing hoodoo there is a good chance that the pen will have laid her eggs. Swan eggs are big. One will make you the equivalent of a six-egg hen omelette and the Gavelwood pair lay four or five eggs in the nest most years. I am pretty sure swans are meant to share the nest-sitting duties, but it rarely seems that way, as the cob will camp out either on the bank or on the river within eyeshot of the nest. I'd like to say there was an easy way to distinguish the sexes, but there is not; they are in plumage identical. From a distance I am hard pushed to tell one from another, but when they are side by side it is easy to see that the female is slighter than the male. Her body tapers like a sailing yacht, whereas the male is more of a tugboat, and the pen neck is much thinner and more delicate. But it is in their demeanour that you can really tell the difference.

One way or another I get to walk past them two or three times a day when they are on the nest, so by now you would think I was part of the landscape, worthy of being ignored. No chance. If he is on the river the moment the cob sees me he stops what he is doing, eyes me up, and if I continue in the direction of the nest, he pushes hard against the current to reach the nest first, where he proceeds to hiss at me even though we are separated by 10 yards of water, reed and bank. If I stop level with the nest the pen remains completely calm, not moving a muscle or uttering a sound, but for the cob this is a goad too far. He'll stretch up in the water, raising his neck and head directly towards the sky to his full height, open his wings and flap them four or five times as hard as he can, then settle back to stare me out. If I head on along the bank he will keep parallel with me in the river until he feels we are a sufficient distance from the nest, at which point he will turn back and with a flourish shake his tail feathers in what I can only assume is the swan version of the V sign to send me on my way.

The eggs take about six weeks to hatch, so this stand-off between the cob and me is a common tableau for all of April right through to the start of the mayfly hatch in mid-May, at which point the nest is abandoned. Sometimes a duck will adopt the nest for a late clutch of eggs, but the swans seem indifferent to this short-term tenant once they have taken to the water with the cygnets. The family group will spend the summer on the river, voraciously feeding on the weed that grows in the river, which is their staple diet. Sometimes they can be annoying parking themselves over a particularly good fishing spot or stripping shallow sections of the weed that would otherwise provide food and cover for the fish. But for the most part they keep their distance and the cob gets less aggressive as the cygnets grow up. By September the young have lost their downy grey plumage and instead have brown/white feathers. It is a tight family group that keeps no more than a few yards apart, the seven of them spending the days grazing together in the river or sometimes assembling on the bank to roost overnight. But the turning of the season, from summer to autumn, marks the end of the compact swan family unity as the father turns on the children.

I don't often have compassion for the juvenile swans, but now it is hard not to as the cob decides it is time for them to leave the family river. For a while they have been learning to fly, practising the take-off up and down the river creating mayhem for the fishermen as they beat the surface with their feet and thrash the water with their wings. Occasionally one of the five will succeed and do a triumphant circle over the meadows, landing back on the water as the king of the river. It is this moment that seems to trigger in the cob a realization that they must go, so he starts corralling the five together, driving them to the far reaches of Gavelwood.

This would be fine except that the cob on the next stretch simply drives them back again. Caught in this pincer movement a couple of the juveniles twig that the game is up, take to the wing and fly off, but for the remainder it is now a relentless persecution that will last the weeks until they leave. The cob is as vicious to his children as he is, or would like to be, to me, swimming at them, flying at them, beating his wings, even mounting and biting them. I'd like to think that the young swans stay because of some misguided parental bond, but equally they may not have mastered flying. This sad discord goes on all day, every day. It's sad because every time the cob drives them away the juveniles drift back to ingratiate themselves, only to be driven off again when they get unacceptably close. Bizarrely the pen seems to be completely indifferent to the whole process, never intervening to help or hinder, watching from a distance until she and the cob are once again the only pair at Gavelwood.

But all this is for the future; for now my obsession is with the amount of water flowing through the river and the amount of weed growing in it. On most rivers the two are accepted facts. Whatever hand nature has dealt you has to be accepted with equanimity. But on chalkstreams you have the chance to play the cards you are dealt to your advantage by managing the flows and nurturing the weed. Yes, at Gavelwood we care for the river weed every bit as much as an obsessive gardener does his lawn, and use the old water-meadow channels to redirect the flows to our advantage.

A bridge is the best place to truly appreciate how very different a chalkstream is to any other river. If you are seeing one for the first time three things will burn in your memory: the gin-clear water, the bright gravel riverbed and the lustrous beds of livid green weed that gently sway from side to side in the current. Pause for a while and your eyes will grow accustomed to the dappling movement of the surface, allowing you to pick out the brown trout that hold steady in the current. Most days you will be distracted by small clouds of insects that dance over the water. Occasionally one will break away from the group and dip down onto the surface, content to drift downstream with the current like a tiny sailboat. The trout you thought was doing nothing will suddenly twitch, swim up to just beneath the glassy surface, open his mouth and suck down the fly, leaving nothing more than a dimple of rings that fade away as he returns to his original spot. This is chalkstream heaven and everything we do is for these moments of perfection.

That weed that fills so much of the riverscape might sound unimportant, more nuisance than asset, but whichever way you cut it a chalkstream would not be a chalkstream without it. Maybe we river folk should find a better word than weed, so as to remove the invasive and insidious implications. In my finer moments I try to keep in mind A. A. Milne, who said, 'Weeds are flowers too, once you get to know them', because get to know them I certainly do. From April to October I'll trim, cosset and nurture the weed as I wade upriver with a hand scythe, cutting it to create that picture of perfection. All that said, not all weed is good weed; some I love, some

I hate, and through the year it will be a constant battle to encourage the former and eliminate the latter.

Weed is good for a whole raft of reasons: cover for the fish, a home for the nymphs, food for all sorts of creatures, and just by its sheer bulk in the river the presence that manages and redirects the flow. Like its neighbours on the banks, river weed has plenty of fantastical names: water crowfoot, water dropwort, fool's watercress, water parsnip and mare's tail are just five of the many more than twenty weeds that grow in the river. Inevitably there is a pecking order, and the king of the river weeds is the water crowfoot *Ranunculus*.

The combination of the rather odd name and the Latin equivalent doesn't help anyone unfamiliar with a chalkstream have an idea what this weed looks like, but think of it as a thick clump of buttercups and you are going in the right direction. It really is just the water equivalent of the same yellow flowers that grow in the meadows, sharing the same Latin prefix as *Ranunculus acris*. The immediate difference in appearance is that the flowers are white instead of yellow and that the fronds grow much, much longer. A meadow buttercup will measure at most 18 inches from root to tip but the river equivalent will stretch out as long as 10 feet in the current.

Just looking at water crowfoot doesn't do justice to its importance; run your finger through the fronds and a whole new world opens up. The first thing you will be struck by is that the fronds are sticky to the touch.

Maybe it feels a little unpleasant, and that's because you are touching microscopic algae. Before you recoil in horror, this algae is the essential food for the tiny nymphs who crawl all over the weed. Open your hand and sure enough there they will be, from nymphs so tiny you will have to squint to make them out to others you will actually feel as they squirm on your palm. And it won't just be nymphs you sieve out; snails and shrimps will be equally populous, all of which use the same weed for food and shelter. Very occasionally a tiny, translucent crayfish will rear up to angrily shake his pincers at you.

Nearly the entire length of the crowfoot is actually floating in the river, secured by a root ball in the gravel at its head. Gently move the long fronds to one side and you'll discover another world that lives in the slack water beneath. There darting around will be a host of tiny fish, nymphs, shrimps and even crayfish who are graduating from the life in the weed itself. This is their little universe, a layer of water maybe 6 inches deep sandwiched between the weed and the riverbed, out of the fast flow that would otherwise sweep them away. Here they can feed on the detritus that gathers around the base or any food that gets washed in. If predators appear it is a short dash to the safety of the weed above. All in all it is no bad place to grow up in.

Water crowfoot thrives because this is a river that gives it what it needs, with fast-flowing, well-oxygenated water. Add to that all the minerals a chalkstream carries and the fact that the sunlight penetrates through the clear water and you have the perfect aquarium. My job as the river keeper, brandishing my scythe and tweaking the sluices, is to maximize all those things to make the crowfoot grow. The more the weed grows (up to a point; more about that in a moment) the greater the amount of habitat for all the creatures to thrive in. In its own way each floating tussock of weed is like the human lung: just a few feet square around the outside, but when the total surface area of the inside is calculated with the thousands of little finger-like alveoli it covers the surface area of a tennis court.

Weed-cutting in some form or another is as old as man's use of the chalkstreams first for power, then irrigation and now fishing. Put simply, the rivers are too healthy for the weed, which if left unchecked will grow and grow until it clogs the watercourse. At this point it deprives itself of the one thing it needs – fast-flowing, well-oxygenated water – so it dies back. The winter floods will clear the dead or dying fronds away and the whole process will resume the following spring. However, if you are a miller who needs to turn his waterwheel in the height of summer, waiting till next year is not an option, so the practice of weed-cutting to keep the main channel open began. It would have been a fairly crude affair, using what we call today a chain scythe, a series of scythe blades bolted together in a row, with a length of chain attached to either end for weight plus a length of rope. One man would stand on one bank, one man on the other, and between them drag the submerged blades slowly upstream, moving the entire thing back and forth between them with a slow sawing motion. Slicing through everything in its path, it was, and is, a crude but effective way of cutting the weed, but no good for delicate trimming or easy discrimination between good and bad weed. However it's a method still used today when the need demands it. On a few sections of Gavelwood, where it is too deep to wade with the hand scythe, it is sometimes the only option.

The advent of the water meadows brought a new level of sophistication to weed-cutting as the drowners took charge. No longer was it crude channel-opening. Now the drowners deployed their acquired skill for water control into the way the weed was cut, to redirect, speed up or slow down the flows. But for all their skills, by the late 1800s it became something of a dogfight as three groups, millers, farmers and anglers, all relied on the chalkstreams for different things. The Victorian newspapers, magazines and books of the time are littered with disputes, sometimes ending up in court. Millers complain that farmers are stealing the water needed to drive the grinding wheels. Farmers complain that the millers are impounding the flows to deprive them of irrigation for the meadows. And the anglers? Well, they rail against both, for either the sudden changes in the water flow or rafts of cut weed creating havoc with their sport. But today the needs of the millers and farmers are mostly a long-distant memory, and we do the weed-cutting for the river and the fish.

Weed-cutting is both a therapeutic and a back-breaking job; viewed from that bridge the sight of the pair of us in waders, abreast of each other, slowly making our way upstream with the silent swish of the scythe through the water, slicing the weed as easily as a hot knife through butter, is as pretty and bucolic as you please. I always look forward to the first hour of the day, the chance to slip into the cool water, look upstream and plot our cutting route. Unhampered by the need to catch fish, with the rhythmic motion of the scythe forcing a slow pace, you sort of melt into the river scenery. After a while the creatures just ignore you.

The water voles keep chewing on their reeds. The moorhens peer out from their nests but don't dash away. The little insects buzz harmlessly around. If the sun is shining the darting nymphs and tiny fish will be everywhere to be seen, glistening in the clear water. Flies will hatch before your very eyes, popping from their nymphal shells, hesitating on the surface whilst their wings dry before flying for the first time. Trout get incredibly bold, so much so that you almost get to step on them before they swim away. Sometimes they bump against your legs, sending a little shiver up your spine. Even the swans will push by with hardly a glance. It is as if having come down to their level you have become one of the creatures; an odd one certainly, but not a dangerous one.

Make no mistake, this is hard work. After an hour the shoulders start to ache. After two hours the arms are tired. The breaks to sharpen the scythe blade grow more frequent as the weariness increases and I will take my time using the stone to hone the cutting edge. But it is easier for us than it was. The old river keepers had to swing a heavy, wooden-handled scythe with a cast-iron blade which was the very implement used to harvest corn before the advent of the combine harvester. Today the handle is light aluminium and the blade thin, tensile steel. But however light it may be, it feels heavier by the hour as the day progresses.

There is some logic to the cutting process; it is not random. There is a plan. Return to that same bridge and look for a pattern in the weed and you should see a chequerboard appear as the cut progresses: the black squares where we just trim the weed, leaving the bulk of it intact, the white squares where we cut it away entirely to leave bright areas of gravel. The idea is that the water will zigzag between the weed squares. Not every river creature likes it fast, not all like it slow. Some like the cover, some the shade. We are trying to build habitats for every creature in the river within every few square yards.

Sometimes we will break with the chequerboard pattern, leaving a bar of weed across the width of the river. This is our version of a living weir, a bank of weed that will hold up the flow to create depth. Other times we will leave a clump of weed untouched to float and grow on the surface. It looks untidy and unkempt but sometimes nature needs it that way. For the water crowfoot it has to break the surface to flower and seed, to propagate the next generation. Some insects appreciate the chance to climb out of the water to hatch or simply sun themselves. Or sometimes we have to take pity on a mallard or moorhen that has chosen a particularly thick raft of weed on which to make her nest, and leave well alone.

At Gavelwood we don't cut the weed every month; in that respect it is a bit like your garden lawn, which grows a lot in the spring and summer, but not much in the winter. On the Evitt we all band together to pick out particular weeks that are set aside for cutting during the fishing season, with one week in each of April, June, July and August. It might sound a little regimented, but the cutting pretty well puts paid to fishing in those weeks, with the water disturbed and the floating detritus getting caught on the flies and lines. It is not quite so bad on the headwaters, where few upstream neighbours are sending down weed, but further downstream where there are 20 miles or more of river above, the rafts of weed are so huge that you might almost park a car on them. Mind you, the fish don't care much; these times are an orgy of food, as all manner of shrimps, nymphs, snails and other goodies are dislodged.

Around mid-April there is definitely a moment when I know that spring is upon us. For a couple of weeks now the valley has been turning from grey-brown to green. The mornings are feeling warmer and the evenings are staying lighter later. The songbirds sing louder and longer. There is a definite intent to the way the voles and mice go about their daily business. The bigger fish are starting to show themselves that bit more, but it is the appearance of one particular insect, the grannom, which heralds the start of spring for me. It is, if you like, that moment when the players run out onto the pitch. Game on.

The grannom would never win a beauty contest; I hate to say this, because it does not do *Brachycentrus subnubilus* any favours in the popularity stakes, but it looks something like a cross between a moth and a cockroach. But trout don't see the grannom that way, and for a few minutes for a few days in the middle of April they will go crazy to devour a hatch of these insects.

Grannom, as sedges or caddis – the latter is the same name, just in Latin – have no real reason to be around this early in the year. There are close to thirty different sedges we'll see around Gavelwood, a small selection of the two hundred-plus species across the British Isles, but they all hatch in June, July or August – clearly the grannom just delights in being different, though as a riverine group, sedges tend to do things differently to other insects. The adults start off in the same way as most, mating and then laying the eggs in the surface that drift down to lodge on the riverbed. The eggs then hatch into larvae that look very similar to a maggot or a small, thin, pale, naked caterpillar. Unlike nymphs, which are good swimmers, larvae are easy pickings for everyone that feeds amongst the gravel, not least the salmon parr that eat them in great quantities between transmuting from fry around their first birthday and heading for sea as smolts. So to survive this onslaught sedge larvae have evolved a remarkable protection strategy that is matched only by our pugilist stickleback.

Essentially they build themselves little cases in which they live until the day that they mutate from larva to pupa, the final underwater stage in their life before they swim to the surface to hatch. In late summer you will see thousands of these abandoned white, bleached cases all over the riverbed or attached to the weed. They are small, maybe an inch to 2 inches long, a quarter of an inch in diameter, and are the shape of a narrow ice-cream cone. The larvae build the case by weaving a sort of silk pillowcase to which attaches little stones, pieces of wood and vegetation. The make-up of the case is determined by what is around at the time – our grannom larva is not that fussy – he just wants to crawl inside to be safe from predators.

Once inside he pokes his head out and feeds as he moves around, dragging the case around with his fore-legs or relying on the current to shift him about. The larva feeds on whatever vegetation is at hand, until the time when he is ready to pupate, the stage when he transforms from a larva to being a fully fledged grannom. But first he has to free himself from the confines of the case, into which he is now effectively wedged. So he secretes a bit more of the silk material which he uses to glue the case to a stone or piece of weed, and once secure begin to eat away the case until free to swim to the surface. Now sporting four wings and half a dozen long antennae, the ascending sedge is very vulnerable in a body designed to fly, not swim. Once at the surface, things are not immediately any better. The sedge will have to scuttle across the water to dry its wings and get flying; it is this commotion and vulnerability that really drives trout wild for them.

It is a strange thing about the grannom hatch, but it is incredibly localized, which might explain why some people might go an entire fishing life without seeing one – you really have to be there. During most days of the season, if I compare notes with friends up and down the valley we'll report much the same hatches on any given day. But with the grannom I can have a monster hatch at one end of the Gavelwood – it never lasts long, maybe half an hour at most – whilst someone just a few hundred yards upriver will not see a single one all day long. But when they appear they bring the river alive as the trout slash the surface at these big morsels of food. The truth is that sedges never get expert at flying, fluttering around and bouncing off the river surface like moths against a lampshade – it should be no surprise that the two are distantly related. Seeing the grannom, whether I am fishing or not, is definitely my spring moment. You can keep your cuckoos, gambolling lambs and Easter bunnies. For me it is most definitely the grannom that marks the start to the fishing season proper, and with it a routine that will keep me busy from now to October.

Fishing could happen every day of the week if we so chose; there are no Sunday prohibitions on fishing in England and Wales as there is in Scotland. However we do hold to what we call a 'rest day'. Not everyone observes them; it is a throwback to gentler days when there were fewer pressures on the river, but we keep to it because, well, some traditions are worth keeping when they are at the same time both useful and pleasurable. For me at least the word 'rest' is something of a misnomer; the day would be better described as a non-fishing day, set aside for all those river jobs that are best done when the fishermen are not around. From an angling perspective the idea is that if you leave the fish to their own devices for almost forty-eight hours, i.e. the last cast on Sunday evening to the first on Tuesday morning, the fishing for the rest of the week will be that much better. Do I or anyone else have any empirical evidence to back up this claim? The answer is no. But even if my head tells me it might be arrant nonsense, in my heart it seems the right thing to do and I cannot believe there are not some benefits from it.

By the time the grannom has put in its fleeting appearance the weekly routine at Gavelwood has fallen into shape. It is really best to look upon the place as a giant garden where you have jobs that are routine, seasonal or annual. In my case routine would be something like mowing the banks, seasonal the weed-cutting, and annual a task like the restoration of North Stream. It is those routine jobs I try to get done on the rest Mondays.

I spend an inordinate amount of my time driving a mower or handling a brush cutter. The banks of Gavelwood end to end, which cover the main river, North Stream and the Brook, reach close to 4 miles. That is a lot of cutting: mown paths along all the banks from which we fish, plus the connecting paths across the meadow. The mowing is easy, the trimming of the fringe, or bankside vegetation, less so. First of all you have to heft a heavy and noisy brush cutter, but more importantly you have to find the right balance between cutting too little and cutting too much. Cut too little and the fishermen can't get to the river. Too much and you remove the plants like hemlock figwort and meadow-rue that make a chalkstream riverbank like no other in the world.

Fortunately for me I have another month before the fringe will need any serious attention, but the grass starts to grow once the early morning frosts become fewer and less harsh. In a mild spring it will be late March, but more commonly sometime in April. The first cut of the year, with that newly mown grass smell, is definitely another of those 'spring is here' moments, but dodging the mower along a bank that has molehills like riverside measles takes the fun out of it.

The number of molehills is really quite stupendous; some areas are so thick with them that we will have to either spade the soil away or run a roller over the ground to flatten them before mowing. The fact is that during the winter when the meadows are at best wet, or at worst flooded, the moles gravitate to the higher, drier banks where they have to dig new tunnels, the numerous hills the spoils from their labour. The moles are really more of a nuisance than anything else. They don't weaken the structure of the banks – the tunnels are too insignificant and close to the surface to have any effect. They don't compete with the voles or mice for territory or food, mostly eating earthworms, grubs and roots. They are certainly not aggressors; if you pick one up you can see why. Under that beautifully soft fur is a body with very little strength in the torso. They sort of flop in your hand. They are pretty well blind as well. All the power is in those front feet, whose appearance belies their power. The feet look like the hands of a human baby – bright pink and crinkled. But each finger is covered in fine white hairs and tipped with iron-hard curved claws for digging. Occasionally I will see one pop his head up from a run, or see the earth moving as he goes about his daily chores, but on the whole they are an invisible presence, bar of course those molehills. Sometimes I find a dead one on the grass; stoats and owls will kill them when they venture above ground, but nobody ever eats them. That diet of worms doesn't make for good eating by others. But the moles are only a nuisance for a few weeks at most, and by mid-May they will have headed back out to the meadows.

In centuries past the water meadows would have been drained of the flooding since mid-March, and the sheep let onto the fields for the 'early bite', which was the growth of grass that the drowning encouraged. Sadly today we just have the remnants of the water-meadow system; the North Stream is the most obvious and viable part of it. Maybe in the years to come we'll revive the entire thing, but even without the drowning the valley is still ahead of the more exposed downland by a few weeks. So the local farmer brings down his flock, plus lambs. It takes me a couple of days to get used to the incessant baa-ing that runs from dawn to dusk, but after a while it is part of the fabric of Gavelwood, and when the sheep leave in mid-May to let us grow a hay crop and graze cattle, the silence will seem just as overwhelming.

Every so often my Monday routine is broken by the arrival of Royston with the trout I ordered way back in the winter. They arrive on the back of a Land Rover in a tank that is fed with oxygen for the 15-mile journey from the fish farm. For reasons that defy any logic, whenever Royston pulls up we'll do exactly the same thing: jump up on the back, lift the lid of the tank and stare down at this splashing, heaving, swirling mass of brown trout whose red spots gleam against the white glow of the interior of the tank. Why do we do it? I know what a trout looks like. Royston certainly does. Just by looking there is no way we can count or measure the fish. I don't know. In truth it is just exciting to see. When it comes to actually putting the fish in the river there is no great science to it; it is more past practice, intuition and guesswork. The Brook we leave unstocked, so the fish will be divided roughly two-thirds in the main river and the remainder in North Stream.

Putting the fish in the Evitt is really quite straightforward, though the North Stream I have had to prepare for by building a fish box, a contraption that will allow us to get to places the Land Rover will not. On the main river we drive up to the bank, Royston jumps up to the tank and dips in a long-handled net to scoop out anywhere between six to ten fish. He hands me the net, I lower it into the water, and within ten seconds they will all be gone. And I mean gone. They don't hang around to acclimatize. They simply shoot off. Vanish. They say owning a yacht is like standing under a shower tearing up fifty-pound notes. Stocking fish feels very much the same.

Actually these fish are more acclimatized than most people imagine; the farm they were reared on takes its water from the same catchment as the Evitt, so the water will have a familiar taste and smell, imperceptible to humans but important to fish. They have also spent the last few months of their time in a stew pond that replicates a river, slowly weaned off a daily diet of fish food. So within a couple of days they will act exactly like their wild counterparts, finding a spot on the river to call his or her own, where they seek sanctuary, rest and food.

The bank along North Stream is starting to grow over the scars of the restoration; the bare earth has a slight tinge of green to it and the trees we trimmed have green sprigs pushing out. The river itself looks everything we hoped after just four months, with a tapestry of shallow gravel sections and then deeper runs. With more light and a healthy flow the crowfoot is starting to grow on the shallows; not enough to require a cut for now, but certainly in June. In the deeper parts starwort has taken hold – a weed that always gets mixed reviews. It certainly looks lovely, lighting up the riverbed in day-glo green, each clump made up of thousands and thousands of tiny leaves, rather like a giant version of the watercress you might grow on your kitchen windowsill. But it is something of a silt trap, and given the chance will out-compete crowfoot in a low-flow year. It is not as invertebrate-friendly either; do the same as before by running your hand through the fronds, and once the silt has cleared the harvest will be one-tenth that of crowfoot. But it grows in places that would otherwise be weedless, so for that reason I like it, and the trout are happy to use it for cover.

The fish box is something I have made from a memory long past when I watched the river keepers of my childhood transporting fish along the river. There is nothing very sophisticated about it; a wooden box 4 feet long, 2 wide and 1 deep. The side panels are fretwork, drilled with dozens of 1-inch-diameter holes to let the water in and out, plus a crude lid in the top and a rope to tow it with. Essentially I don my waders, step down into North Stream with it, and allow it to submerge, flip the lid and pour in as many fish as it will take that are passed to me in a net from the transporter on the bank. Then I make my way down the stream, stopping every so often to remove a fish by hand and release it.

The last one in the box always takes an age to catch, but it's worth the effort. Brown trout are a sedentary bunch and in half an hour with the box I will spread them out along the length of North Stream in what would otherwise take weeks if we just put the trout all in the one spot.

So by the time the end of April looms the die is cast; the river is full, the weed is cut, the banks are trimmed and the fish are stocked. Everything up to now has been the prelude for May, the single most exciting month in the chalkstream calendar, the month when nature lays on a show the like of which you will only ever see in these few river valleys. Mayfly time is just around the corner.

8
MAYDAY

THE ANGLING WORLD is full of contradictions; not least the fact that May 1st is regarded by many as the traditional opening day of the chalkstream season, despite the obvious truth that the proper and legal start is April 3rd. A bit of me rails against the inconsistency of it all, but on the other hand the quirkiness is part of the charm, so if people want to let me have the run of the river for a month, who am I to complain? Actually, though I hate to say it, there is a certain logic to a May start. Not everyone is prepared to freeze their fingers to the bone or wait for days on the off chance for thirty minutes of grannom action. May is the month when the river wakes up as the days get longer and warmer, when you can almost taste the difference between April 1st and May 1st – you can certainly see and hear it. Pretty well everyone is up and about. Every bird is sitting on eggs close to hatching or shepherding a brood along the river. The otter, water vole and field mouse work all hours, day and night gathering food to take back to the mewling infants. The trout are on their game and a river apparently devoid of fish a month ago now has them in every spot.

For the people it is the Drowners House that has become the hub of Gavelwood each morning, the place where the anglers and helpers gather before the serious business of the day, be it fishing or river work. 'I'll meet you at Drowners' has become one of those stock phrases instead of 'I'll see you later.' In my mind I had hatched a plan to build a new fishing cabin, but over the winter as Drowners became our default location when the weather was bad or we needed a point to meet at, I turned my mind to restoring it.

The first time I saw Drowners back in the summer it was a mess. It sits directly over Katherine's Brook, straddling the stream that races under the building, which was open inside the house – the wooden floor that covered it had rotted away years ago. Three stout cross-beams remained, and it was clear from the fresh spraints that the most recent inhabitants were the otters, who clearly found it the ideal spot to rest and gorge on a trout or two. In places the thatched roof had collapsed to let blue sky show through, but otherwise the building was as sound as the day it was built.

Drowners is old; at least four centuries, with thick red-brick walls and rain-worn Portland stone corner-stones. It is more or less 15 feet square, with a thatched roof supported by oak rafters that rise to an apex from which hangs a storm lantern on a chain that is just about low enough for the unwary to hit their heads on. It had survived because it was built to survive; flood-plain buildings are designed that way. The foundations are huge lumps of stone that prevent the stream washing away the ground. The walls are deep, four bricks thick, tapering ever so slightly from base to head height. The mortar is lime, which flexes as the building moves on ground that expands when sodden in winter and shrinks when drying in summer. Outside the house there is a leat, or man-made stream. This was dug to divert the heavy flows coming down the Brook in times of flood away from the house. All in all, Drowners was built to last, but sadly not always for comfort.

Besides the obvious, a river running through it, there are two things that will strike you at once as you enter a drowners house for the first time. First, the lack of windows. They simply don't have any. Second, the damp. The wooden floor (when there is one), the walls and the air are all damp. Winter, spring, summer, it is always the same, as is the temperature. A warm 51°F when the frost is on the ground, a cool 51°F when the sun blazes down. The latter was of no interest to the original occupants of the house, the drowners, who only ever occupied it from January to March, when the flooding of the water meadows required them to be there around the clock. Today is another story, and the restoration posed a question: do we restore it to what it was even though its original purpose is defunct, or do we change it to something useful for today? The restoration was going to involve three things: a new oak floor, two new oak doors and a major overhaul of the thatch, which is unusually sedge reed, as all the thatched houses in the village are either Norfolk reed or straw. But since the sedge grows prolifically around Gavelwood it was at the same time local and, even better, free. The real problem was the windows, or rather the lack of them. Should we knock holes in the walls to create them? In the end the decision made itself.

The roof seemed the natural place to start, and with the roofing material all around us we got to work. Sedge is one of those unspoken staples of the water meadows; until you have a purpose for it and start looking you don't realize that it is absolutely everywhere: in the margins, on the banks and across the meadows. It grows in clumps; it grows in swathes and most impressively tussocks with thin, sage-green, razor-sharp blades sprouting from the head. These grass-like blades are truly razor-sharp – pull at a handful and your palm will be shredded by incisions as painful as any paper cut.

The tussocks look like something out of a sci-fi movie, growing in clusters, huddled along the bank where they like their roots in water year round. Each is the size of an upended wine barrel; apparently brown and dead all the way up from the ground to the head, where the grass sprouts up before tumbling down to create an all-round fringe that almost reaches the ground. These tussocks can outlive trees; in places I have fished since childhood the same ones still grow, the only difference today being that I am now tall enough to perch my bum on them as a comfortable resting spot whilst waiting for a fish to show. The tussocks, for all their freakish appearance, have something of a reputation as a refuge habitat, creating a microclimate, if you want to call it that, where the insects and tiny mammals can survive the extremes of flood and drought. It is said that if you shake out a tussock over sixty different species of invertebrate will fall out. I can't swear it is true but I can confirm, having sat on quite a few, that some of the inhabitants, probably ants, have a nasty bite.

In recent years there has been something of a dust-up between ecologists and river keepers on how to manage these tussocks. You may have seen evidence of the traditional practice while driving through the Highlands or the west coast of Ireland, where they proliferate and are managed by an annual burning. And burn they do. The air-dried blades catch easily, popping as the flames jump from tussock head to tussock head leaving a smoking landscape of bald, blackened sedge heads. It is certainly an effective way to clear the debris and there is no doubt that the tussocks grow with renewed vigour each subsequent year. However, it is something of a scorched-earth fate for the poor insects that mostly perish in the flames.

So in deference to them, we trim the tussocks with a hedge cutter, leaving the debris on the ground for a few weeks before finally raking it away.

The tussock trimmings are not what we are after for the roof; it is the sedge that grows in swathes in the wet sections of the meadows. During the summer and autumn the cattle leave it alone; the ground is too wet for them to easily forage, and they don't eat it for the same reasons you should not pull at it. So by December, entirely by an accident of nature, a crop is there to harvest. Cutting is simple. The same scythes we use for the river weed are given an extra fierce edge to the blade and we bend our backs into the base of the sedge reed beds. It is a bit like cutting a wheat field, as three abreast we scythe the pale green blades to the ground. Amongst the harvest is plenty of dead matter, what a thatcher would call 'litter', which we leave behind as we gather and bind the sedge into bundles that we stack crossways like a squat Jenga tower to dry out.

By March they were dry, ready for patching the holes in the roof. In truth Drowners needs stripping back for a complete rethatch, but for now make-do will do. I am no thatcher, and if we were starting from scratch I'd be lost. But I have seen it done enough to know the principles, and by following the pattern of what remained the bundles were tamped into place, compressed by hand and finally held in place by split hazel pegs bent into an inverted U shape. Onto the floor went freshly sawn oak boards, still wet with sap, so recent was the felling of the tree. Dried boards were no use to us; the damp interior would soon have them expanding and twisting. The doors were made of the same, the hinges rescued from the skeletal remains of the old doors.

While trying to admire our handiwork we found out why windows were a must. With the holes in the roof repaired and the doors closed, the interior was pitch-black and not a little spooky as the rush of the water beneath filled the darkness. The drowners must have been a hardy lot, but we were not, so two windows were promptly knocked into the south and east walls to complete the restoration. With the addition of a table, some benches and the storm lantern, Drowners was complete with the exception of one item – a catch record book.

Record books come in all shapes and sizes. Big leather-bound ledgers embossed with the name of the fishery. Card indexes. Cloth-wrapped diaries. Bespoke books, each page laid out to record the date, weather, number and size of fish caught, flies used, and an 'other comments' column for embellishments and observations. Scrappy school exercise books doctored to do the same at a fraction of the cost. One famous River Test beat has the records inked on the oak walls, including a 1944 D-Day catch return with the comment, 'A. Hitler be damned!' However, make no mistake, whatever the vellum, whatever the binding, these are important tomes. On a salmon river it is all about red-toothed capitalism, on a chalkstream rather more about sedate entomology.

The value of a salmon river resides entirely in the number of fish caught; a productive beat by a busy road will always sell for more than an idyllic but less productive beat. The different weeks of the season are priced according to the historic catches. Rent a prime week and it will be twenty times more expensive than a low-season week. Estate agents assessing a beat for sale or probate will use the average catch over the past ten years to arrive at their valuation, and with some changing hands for millions the value of each fish suddenly becomes a critical multiplier. Salmon river owners recognize this, and when the chips are down they don't suffer fools, or more pertinently bad anglers, gladly. If you think your money alone is enough to secure the best week on the best beat of the best river, you'd be wrong. You need to deliver a good catch return, and if you don't, well, suddenly that week will become 'unavailable'. On the other hand if you are a good salmon angler who can magic fish apparently out of thin air, your inbox will be thick with invitations for the most amazing fishing for absolutely nothing. All round it is a fair exchange.

Chalkstreams are very different; the numbers matter less and the tone more. The mark of a successful river is rising fish. It is great to see fish, it is even exciting to catch them on a sunken nymph, but the money shot is the fish that takes the insect from the surface. If you think of most angling as monochrome, dry fly-fishing is the full colour, HD/3D model. That moment when a dimple radiates across the river surface just ahead of you as a trout sucks down a fly is truly every bit as good as eyes meeting across a crowded room. Fish that don't rise are as pointless as pheasants that don't fly when dry fly-fishing is the game, so the most depressing remark in the comments column of the book is always, 'Didn't see a rise all day'. On the other hand a comment along the lines of, 'Fish rising all day but could not work out what they were taking' brings out the alpha male in fly-fishermen who, even though they would never admit it, secretly expect to succeed in spades where the author of the comment has failed.

So watch any angler arrive at a fishery and you will observe a pattern. After the usual pleasantries – the journey, the weather past, present and future and an update from the river keeper of his latest travails – the angler will peel away into the cabin intent on examining the catch record book, scouring the latest entries for clues to what flies to use or which parts of the river are fishing best. It is a ritual that is hard to avoid, though what it does for your state of mind is open to debate. If the diary reports a cracking few days you are pitched into both envy, which feels mean, and mindless optimism that the winning streak will continue. If the reports are dire you get no clues as to what flies are working and are pitched the other way into a slough of despair that you have chosen a bad day to fish.

For myself I'm a less than enthusiastic reader of record books, for all of the above reasons when it comes to the fish caught; fishermen are not liars but they are mostly natural optimists who prefer, like gamblers, to celebrate their successes above mourning their failures. Likewise since catch and release has become a common practice, the number and size of the fish caught has inexplicably risen now that the return is estimated rather than weighed . . . But when it comes to the comments section and brief remarks like, 'Good hatch of BWOs [blue-winged olives]', my own little antennae start to twitch. For a successful chalkstream day is all about the hatches, and around Mayday the hawthorn hatch is the one we hanker after.

There are certain insects that don't really belong on a chalkstream, but are intrinsic to the hatch calendar. They are collectively known as terrestrials – as the name indicates, bugs that live on the land rather than the water. Daddy-long-legs, or crane flies, are the most easily recognizable members of this group, which includes flying ants, beetles and the hawthorn fly. It is the misfortune of these insects to be blown onto the river where, ever the opportunists, the trout will go mad for them at particular times of year. Now why trout have such a predilection for things that have no place on the river, which only appear for a few days each year, is as much a mystery as why cats like fish. But anglers, like the trout they seek, are opportunists in equal measure, and the emergence of the hawthorn fly is the clarion call to head for the river.

There is some debate as to why the hawthorn fly is so called. Some say it is because it lays its eggs in the roots of the hawthorn bush, others that the hatch coincides with the flowering of the bush of the same name in late April, early May. I have no idea which is definitively true (for the record I err on the side of the bush theory), but whichever is accurate they are a freaky, fun way to celebrate Mayday. They are freaky because, well, they look like freaks. There is no other fly in the entire year that looks remotely like a hawthorn. To start with it is jet, jet-black with a pair of long, hairy legs that hang down like a lanky undercarriage. Secondly it is instantly recognizable, as it looks a bit like the common housefly; the two are distantly related, coming from the Diptera order of true flies. However the hawthorn appears to have been on an intensive course of steroids that gives it a bulky appearance, making it more than twice the size of its domestic counterpart. And like most over-muscled creatures it lacks nimbleness. Hawthorn are easy to pluck from the air with your hand as they fly past, their direction of travel dictated by the wind rather than any purpose on their part. Their life on the wing is short, just a week during which they hatch, mate and then lay their eggs in the soil from which the fly will emerge from the larval state fifty-one weeks later.

Hawthorn are fun because they are something of a circus act: the show that hits town for a few days creating entertainment and mayhem before it moves on for another year, leaving a void in its place. They are easy to spot as they swarm together as part of the mating ritual, keeping close to bushes and undergrowth in the warmth of the day. Clearly hawthorn are aware of their aeronautic shortcomings. Unlike the mayfly that will rise to 20 or 30 feet in the mating dance, hawthorn rarely go over human head height. But there are always a few that peel away or get caught by a sudden gust, and the observant angler will track the progress of any that come in the direction of the river.

Once you see them coming your way there is a strange inevitability to their fate, like the hapless canoeist inexorably being carried towards the waterfall. I open my fly box and tie on one of the hawthorn imitations, which have spent as much time in a closed compartment as the larva has in the ground. At this point I have two choices: cast to where I know a fish will be in the confident knowledge that the hawthorn is the dish of the day, or wait a few minutes for the first of the clowns to tumble onto the water, bringing the fish to the surface.

Now here is a strange thing about fish and fly-fishing in general. If I do the former, nine times out of ten the fish will ignore my imitation. OK, you might say, fair enough. After all it is a full twelve months since the fish last saw a hawthorn (or never in the case of the yearlings and stocked fish), so he'll need to see a few to wise up to them. I like this theory. It makes sense. Until, of course, the very first hawthorn plops onto the surface and wham!, up comes a trout. Follow up with your imitation and wham!, you'll have a fish on. Confounding, but such is the endless fascination of fishing.

It is this wham, bam, thank you ma'am approach that reignites the flames in the heart of every angler at the start of the new season. Sure it is good to tie on tiny flies to imitate obscure hatches, but that is more for the balmy days of summer. Early season, when it is still a little chilly with gusty days, the casting action is a bit rusty and the fish hard to spot, this circus act is the best entertainment ever, or so we think at the time. We all know in our hearts it is really a bit too easy, but who can resist? And then as suddenly as they arrived the hawthorn are gone. They don't taper off, they stop dead. One day they are there, the next gone for another year. Suddenly the river seems empty. Disconsolate anglers stare at the flat calm or peer hopefully across the meadows in the vain hope of more hawthorn blowing their way. But no such thing. Even the fish take a break. The circus has left town. We are sad to see it go, but the emptiness will be short-lived. In a few days the first mayfly will appear.

9
THE MAYFLY

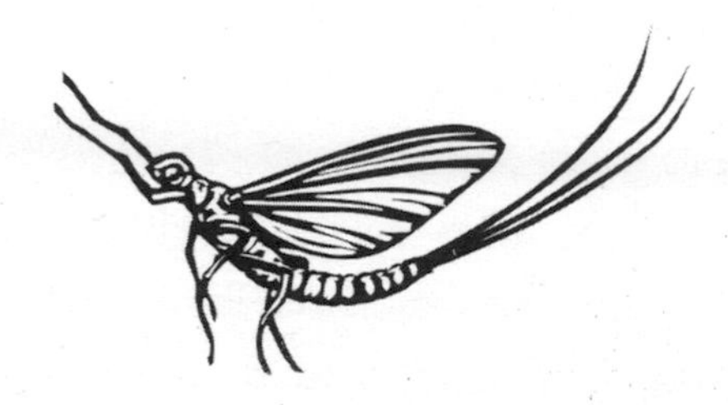

WHICHEVER WAY YOU cut it, the mayfly, *Ephemera danica*, is the iconic chalkstream insect. It defines the most exciting moments in the fishing calendar. It is the proof that the chalkstreams are the most perfect rivers ever created. It is the moment when sometimes you have to step back, shut out the rest of the world and watch in awe to simply accept that nature is utterly and completely amazing.

The mayfly will live for just twenty-four hours at most, perfectly adapted by millions of years of evolution to execute the perpetuation of the species over that single day. There is no time to be wasted. The mayfly does not even have a stomach. It does not live long enough to need to eat. Hatching in their tens of thousands, from a distance the huge clouds of insects look like gunsmoke drifting beside the river. Get up close and an elaborate mating dance is taking place, at the same time both graceful and frenetic as two from the thousands pair up to consummate in a moment what has been two years in the making. But the full story of those two years is not all played out in the bright sunshine, along the verdant banks of the river, but rather in the dark, muddy recesses out of sight of every person and most other creatures, a far cry from those romantic last twenty-four hours.

There is a slight irony to the work we have done on North Stream and the vision of the perfect chalkstream we seek to attain, given that the bright gravel is favoured over silt or mud, when the latter is home to the mayfly for almost all of its life, and without the silt the chalkstream would be devoid of its trademark hatch. However, that said there will always be enough diversity in the different habitats of Gavelwood with the main river, stream, brook and carriers, to preserve the silty bottom the mayfly nymph craves. The only parts of a river that have a problem are the headwaters close to the source of the Evitt where the fast, shallow water strips away any silt build-up, or at worst it dries up in the summer. Unsurprisingly, the higher sections of any chalkstream will have a sparse mayfly hatch at best, the few nymphs that do exist preferring to drift downstream until they find a better and safer home.

Fishermen are often dangerously imprecise about the words they use to describe particular insects or hatch, but the word 'mayfly' has a precise meaning on the chalkstreams that exists nowhere else in the world. This causes great confusion with anglers from overseas, and considerable annoyance to entomologists who don't like to see their taxonomy bastardized. But even though tradition trumps the science of classification on the rivers of southern England the taxonomy still insists that there are over two thousand different types of mayfly worldwide, with around fifty in the British Isles. They are all of the order Ephemeroptera, essentially small up-winged insects that only live for a few days after hatching. However, at Gavelwood when we say 'mayfly' we mean only one of the two thousand, *Ephemera danica*, that hatches for two or three weeks across May and June. Anyone else, anywhere else in the world, could be referring to any of the other one thousand nine hundred and ninety-nine or so others of the Ephemeroptera order that hatch in May . . . or so you would think. Fortunately, chalkstream anglers are not alone in abusing the language, as mayflies, up-winged insects, hatch not only in May but in each and every month of the year. In Victorian times they were collectively referred to as 'dayflies', which if you think about it, better describes their lifestyle.

But whatever the name the cycle starts with the tiny eggs, something less than a millimetre in diameter, being laid in the surface of the water and then drifting to the riverbed. The sticky casing helps the eggs attach to the bottom, from which they hatch into tiny nymphs that in turn head straight for the safety of the nearest silt bed. Here they burrow, and this will be the mayfly's home for the next two years. From the angler's point of view the mayfly nymph now becomes something of a non-person, lurking out of sight. Other mayflies are more obliging: the nymph of the large dark olive that likes to hatch in March and April is a darter, spending his time underwater moving from stone to stone in short bursts that give the trout something to eat and the angler something to imitate. But with two years ahead of him, growing by frequently moulting his outer skin, maybe hiding in the dark recesses of the chalkstream, is the best means of survival for the mayfly nymph.

It sounds a strange thing to say, but the quality of the silt is critical to the survival chances of the nymph. Part of the reason to restore North Stream by flushing it through was that the accumulated mud was, to coin a phrase, the wrong type of silt. Out with the bad and in with the new; between the fast gravel sections, the slower deeper water was the ideal place for new silt beds to accumulate, with fresh material washing downstream. Nymphs have to breathe through their tiny gills, not unlike fish in that respect, and the fetid, compressed mud of pre-restoration North Stream would kill them. What the nymphs need is loose, sandy silt that is easy to tunnel through and is constantly percolated by the oxygen-rich water, combined with masses of vegetable matter like dead leaves, reeds and weed roots for the nymphs to chomp on. Six months on from the day we opened Portland Hatches, the silt beds had begun to establish, but they were too new to hold many mayfly nymphs other than some that might have been washed down in the winter floods. I had plans to change this, but I needed the help of *Ephemera danica* and some willing volunteers in the weeks to come.

There comes a time in the life of every mayfly nymph when he says goodbye to his dark tunnels, pushing out of the silt and swimming into the open stream in preparation for his or her date with destiny. Not the best of swimmers – he is no agile darter for sure – the mayfly nymph becomes fair game for the trout who switch on to them, consuming them in their thousands once they have started to leave the safety of the silt beds in late April. Common sense tells you that the nymphs should head straight for the surface, hatch and get on with it. Why hang around and risk being eaten? But nature has a plan, insisting that they gather in their legions beneath the surface so that when the trumpet sounds they will hatch as one.

Of course there is always the occasional nymph that does the commonsense thing, hatching as a sad singleton weeks ahead of schedule. In a way it is a good plan to at least escape the jaws of a hungry trout, but as a way of finding a mate it is a hopeless strategy, and scant reward for two years spent in the silt. Whenever I see a solitary one being buffeted hither and thither on a blustery April day part of me feels happy for the affirmation that the cycle of life is turning full circle, but another part feels sad that this is ultimately a life that will end in failure, usually snuffed out by an eager trout. Fortunately for the perpetuation of the species these outriders are few and far between, the hatch-day timing dictated by a force of nature that the mayfly nymphs are conditioned to obey, greeting the world on a date that anglers set their calendar to.

Nobody entirely knows why the mayfly is able to time its arrival, the second week in May, so precisely. This is no random event – go back through the catch record books of long-established fisheries for a century or more and you will find that clarion cry 'The mayfly is up!' inscribed on a date close to 10 May each and every year. It is perplexing when you think about it. Surely an incredibly cold or wet spell or some other weather extreme in the months leading up to May would shift the date? It does for nesting birds, trees, plants and most other creatures, but not the mayfly. The daffodils in your garden can be anything up to a month late. Maybe it is the weather itself in May? But that doesn't stack up against my days on the river where the hatch has happened in everything from extreme gales where we never even put a rod up, to days when we returned home with sunburn. Maybe it is the water temperature, but the remarkable consistency of a chalkstream in this respect suggests otherwise. In the end the only constant that straddles the records I see is the length of the day and the light intensity that goes with it; somewhere in that little nymph shell it tells him the time is now.

Arrive at a river on hatch-day morning and you will surely be disappointed. Where, you will reasonably ask, are the promised clouds of mayfly? The river will be flat calm. You will probably be tempted to use the word dead. The fish won't be much bothered, idly holding station, oblivious to all. As a fisherman you compute that the fish will surely have a memory of a juicy mayfly, so you tie one on. Each cast you make is ignored with increasing indifference until a bad one slaps the water; now at last the fish despairs of your cack-handed effort and swims for cover under the weed. Chastened, you engage your fishing intelligence. Maybe a small fly is called for? A snack. A little *amuse-bouche* before the dish of the day. Out comes a tiny Dark Olive, but the outcome is still angler nil, fish one. You retire if not hurt, then a little miffed. You'd been told this is Duffers Fortnight, the moment in the season when even the most inept angler is guaranteed success. Is it you or is it the fish? In fact it is neither, it is the mayfly.

Whatever anyone tells you, fish do have a memory, and though it certainly does not put them in Mensa territory it is considerably longer than seven seconds. They know deep in their fishy brains that if they wait the hatch will come to them with a feast of plenty, so until then why expend unnecessary energy? And start it will, usually sometime in the late morning, when you'll catch the first fluttering mayfly of the day out of the corner of your eye. If the day is warm with a little bit of breeze, the delicate insect, about the size and colour of a dandelion seed head, will rise quickly into the sky and before you really realize it was ever there, be gone. Blink and you miss it possibly, but once the hatch starts in earnest you will not miss the mayflies that follow. At first you will be able to nod your head at each individual mayfly as it rises off the surface to take flight, even to the point that you start counting them. But soon the intensity of the hatch will increase so that you hardly know where to look, until the first slashing rise of a greedy trout brings you abruptly back to the river surface.

It takes that first rise to wake the river up. Whether it is that one trout that spurs on the others or whether the conditions are suddenly just right for them all I do not know, but one rise begets another, then another, then another, until you barely know where to look as the river surface churns with fish rising. And the trout are not uniform in their approach to the feast. Some take up position under the food line, the point where the current concentrates the flow of insects on the surface, purposely rising metronomically, carefully selecting from the mayfly parade that drifts above. Sometimes you can almost predict the feeding pattern as they rise for one, submerge to chew it down, and then rise again for the next a minute later. Others abandon their usual haunt, scooting around the river, gulping down any mayfly that catches their eye, ignoring the rule that trout feed facing upstream as they chase mayflies sideways, crossways and downstream. Some throw caution to the wind, leaping into the air to grab the mayfly in flight, crashing down causing a commotion that appears to make absolutely no difference to the other fish who continue to feed oblivious.

That feeding pattern of the trout is often dictated by the weather, as the freshly hatched mayfly drift down the river on the water surface like little sailboats, making ready for that maiden flight. On a damp day it takes a few minutes for the wings to dry; easy pickings, so more of the metronomic trout. On a warm day it can be just a matter of a few seconds, so the trout are more frenetic. But whether the conditions dictate that it is seconds or minutes on the water, the sheer fact that mayfly go so fast from being a nymph in the river to reach the surface, break out of the hard shell and emerge ready to fly is amazing in itself. And fly they do. There is none of the struggling of the inept sedge or wind-borne randomness of the hawthorn. Mayfly, from the moment they first stretch their wings, fly with a purpose. For the male *Ephemera danica* the flight is short-lived; up in the air, do a few turns and then head for the nearest bush or tree to hang upside down under a leaf. The first hatchers of the day will be the males, who at this point are not much use to anyone, as they are sexually immature, what the entomologists call sub-imago. But not for long (time is always of the essence for the mayfly), as he performs one final moult to emerge 'imago': bright, shiny and with those large transparent wings ready for the mating dance later in the day.

Back on the river as the day progresses the hatching will become more intense, to the point where, for maybe no more than half an hour, it will reach a crescendo. The river surface will grow thick with mayfly, dozens covering every square yard, the legions constantly replenished as for every one that takes flight three more pop up from the water beneath. The trout gorge themselves silly as the mayfly keep coming, but even a trout has its limit, and as the hatch dies away the feeding will become desultory, so that by mid-afternoon, with all the mayfly headed for cover in the meadows, the river is quiet again. But this is not the end by any means; for fish and mayfly the best is yet to come.

The mayfly mating dance is one of the defining moments in the chalkstream year; whether you are fishing or just working in the meadow, the moment to stop and gawp. Like so much to do with the life of the mayfly, it is like a magic trick – one minute nothing, the next the reveal to produce a display with few equals. Walk the banks and meadows in late afternoon and you would be forgiven for thinking the mayfly hatch is a mythical event. But stop for a moment, peer under the leaves of any bush, tree, plant and even thick grass, and there they will be. Thousands and thousands, hanging upside down, half a dozen to a big leaf, ready for the first and final dance.

As mid- turns to late afternoon the occasional insect will lift off, fluttering around as if to say, 'Hey girls, here I am.' It is the males that are the first on the dance floor at this disco, but being first is not always the wisest decision. The mayfly are soon joined in the air by the swallows and house martins, twisting and turning in the air to snatch every mayfly that takes flight, grabbing them even milliseconds into flight when they have barely had a chance to leave their perch. It is the misfortune of the mayfly that these late spring arrivals from southern Europe and north Africa like the very same places they do: open meadows, a few trees and of course, water. But the mayfly have a way of dealing with the threat – mass flight. Suddenly, as if to protest at the death of the outriders, the mayfly rise up in their tens of thousands. The empty air above the meadows is filled with insects. There are so many that the blue sky on the far horizon looks as if a gauze has been drawn across it. Strangely these huge clouds of mayfly seem to intimidate the birds, which back off, patrolling around the edges to pick off the stragglers.

Make no mistake, the clouds of *danica* are thick; I guess the birds don't fly through them because it would be akin to invading a hailstorm, peppered by bodies you could not avoid hitting. But what looks like a chaotic morass from a distance is in fact a closely choreographed mating ritual as the mayflies gather in columns, flying up and then drifting down. It is as if each mayfly rides in its own individual elevator, shooting up to the top floor, then gliding back down to the ground floor, before hitting the up button again. Up and down they go, anything from a few feet above the ground to

30 feet in the air, until one breaks the pattern, darting horizontally to grasp another in midair, where the coupling takes place as they continue to fly up and down.

The mating is necessarily short-lived; flying in unison is hard work, and it is not helped when a female attracts two or three suitors simultaneously. Once done, that is the end of the line for the male mayfly; all that is left is a return to the bushes and death. For the female the final act is egg-laying on the river. If mating has happened very late in the day she'll wait until the following morning, but for the vast majority they have maybe fifteen minutes of life left to live. They'll find the river, dip down onto the surface, and after exhausting themselves in the act of squeezing out thousands of eggs, collapse in the film to drown. By dusk the river is quiet again; the last of the mayfly have died on the water and the replete trout lie placid near the riverbed, digesting a huge meal on what is likely to have been one of the heaviest feeding days of their lives. But for us at Gavelwood on this particular day there is one last task; the gathering of the mayfly to repopulate North Stream. Of course we could leave it to nature – I am sure that over a period of years the mayfly would find their way to North Stream when a fierce spring gale whisks them across the meadow to alight on a new home – but I am too impatient to wait on the vagaries of the weather.

I'd heard of this happening more by rumour than fact. On the face of it, it is a fantastic feat to pull off. How do you cope with such a delicate, ephemeral creature, transport it miles from one river to the next? But there are plenty of stories of barren rivers being repopulated after years of no mayfly, or the stock revitalized by the introduction of adults from a different part of the river when years of interbreeding have seen a gradual decline as the lack of diversity reduces the ability of the mayfly population to defend itself against diseases. In my mind I had some fanciful ideas. Sieving the river surface to capture the eggs before they drifted to the riverbed. Elaborate nymph traps. Giant butterfly-style nets to capture the mayflies in the air. Of course none of this is necessary, because the mayfly, by virtue of its leaf-hanging habits, unwittingly simplifies the difficult.

By the end of the day not every female has laid her eggs; impregnated, yes, but spent of ova, no. Some have simply run out of time, others made a start but still have more to do. One way or another they have enough instinct to know that tomorrow's dawn will open the door to their destiny. So they head to the leaves of the bushes and trees along the bank for the night, sharing the space with males who are likewise waiting their turn tomorrow. To say these insects become inert once dusk creeps up rather understates their torpidness. You can shout at them, waggle the branches or shake the leaves, but will they move or even flicker? Not a chance. If I didn't know better I would swear they are all dead, but for us mayfly gatherers it makes the task very easy as we gently lift the female mayflies by the tips of their folded wings, ease them off the leaves and lower them into a cardboard box by the dozen. They don't protest, they don't flutter. They seem utterly uncaring that they have been rehoused.

I don't know why, but I feel impelled to include a few males in the harvest, who are easy to spot when side by side with the gravid female who is larger, brighter, and has a butter-yellow body compared with the beige of the male. There is no science as to how many females should be gathered, so we keep at it until the gloom makes it too hard to distinguish the sexes. Peering into the boxes where they cling motionless to the sides, I guess there must be hundreds of mayflies all told, maybe over a thousand. Multiply that by the thousands of eggs that each carries and surely this will be enough for the repopulation to succeed? We have a bit of a debate as to whether to move the boxes to the North Stream that evening or in the morning. The only possible reason not to might be a difference in temperature in the different parts of the river valley – hardly likely to have an effect, but after all this work the risk seems unnecessary, so we leave mayflies in the boxes under the same bushes they were gathered from. The only thing of which I am certain is that I want to make the journey from the mother river to North Stream soon after dawn, whilst the mayflies are still dormant in the chill of early morning.

The cardboard of the boxes, damp from the early morning dew, was easy to slice along each side. Once opened up, the boxes were laid flat in the shade of the bushes alongside North Stream, the serried ranks of the mayflies still comatose, unaware of the overnight activity and their new home. It might have been tempting to shake the boxes out over the water to deposit the mayflies directly on the surface, but over the years I have noticed that this would be a long way from how the females conduct themselves on this all-important day.

Nothing really happens until the morning starts to warm; a bit of rain or wind doesn't particularly matter. It is the temperature that counts. In human terms it would be that time when you shrug off the extra layer you put on before leaving home. One by one the mayflies start to rise from their cardboard perch, lifting up a few feet before unerringly heading for the water. It is uncanny, but none of them seem to have the slightest doubt in which direction the river lies, and once they reach it there is only a moment of hesitation, or maybe reconnaissance, before they drop onto the surface. There are two styles of egg-laying: the passive drift and the flying dip. The drifters land on the surface, where they remain, pushing out a long string of eggs that is split into shorter sections by the movement of the current. The dippers push out the string in the air whilst flying, dipping briefly onto the water and using the surface tension to break the string. They bounce up and down, two, three, four, five times, until they too come to rest on the water, the supply of eggs exhausted, the end of their lives just minutes away.

On a normal day I would watch this activity with a different eye, willing the trout to rise and snaffle down the egg-laden mayflies. But today I feel very different. I haven't captured all these females for them to be eaten before they have had a chance to lay their eggs. My hope is that the trout on North Stream, who have mostly never seen a mayfly, will take a while to catch on to the potential feast, but that proves to be a vain hope as the first of the drifters land on the stream only to be consumed within a few yards of touchdown. As the number of mayflies on the water redoubles the safety in numbers equation kicks in, and plenty get the 30 or 40 yards of drift they need to complete egg-laying. Morning turns to a warm afternoon, quickly thinning out the ranks of mayflies on the cardboard as the rise in temperature encourages them onto the stream in ever-greater numbers. The dippers start to outnumber the drifters, much to the annoyance of the more impatient trout who, having to expend more effort than absolutely necessary, start to throw themselves in the air or accelerate up from the deep to snatch a mayfly at the moment of take-off.

Not all the trout are this exuberant; some are wise to the ways of the mayfly. Way down North Stream, almost out of sight of the release point, a group of trout have tucked themselves into the side of the stream under the back eddies. With their dorsal fins out of the water they gently circle like sharks, from time to time lifting their head and mouth to gulp something down. From the distance it is impossible to say what. Nothing shows as sitting up on the surface. But get closer and the water is littered with mayfly corpses, their bodies spreadeagled, held flat by the surface tension. Death comes quickly to the female mayfly. Some simply collapse. Others put up a fight, pulsating in the surface as if in the throes of an epileptic fit. Whether they resist the inevitable or not, the end soon comes to the spent female. But the purpose of their life is fulfilled. The eggs are laid, starting to settle on a riverbed primed for this very purpose. Two years from now, almost to the day, we will know if the repopulation has worked.

To the uninitiated, mayfly sounds a rather innocuous word; an insect that hatches in May – what's the big deal? An official definition that reads, 'a slender insect with delicate membranous wings having an aquatic larval stage and terrestrial adult stage usually lasting less than two days', I don't find a great deal more help. Yawn. This is more likely to send you to sleep than fishing, but scratch beneath the bland words and you'll discover an amazing natural phenomenon that excites fish, fishermen and river creatures like no other.

Fly-fishing is all about taking a hook and embellishing it with thread, feather and fine wire to make a passable imitation of a natural insect that you float on the surface to fool the hungry trout looking up from the river below. Believe me, there are many easier and more productive ways to catch fish, but the delight of fly-fishing lies somewhere deep in our psyche. Outwitting a wild creature that lives on its wits is immensely satisfying, though I sometimes wonder why this is when you consider that the IQ of any fish is probably in negative territory.

Fly hatches are everyday events on the chalkstreams, and for that matter most rivers, so why is it that the mayfly excites like no other? Well, everyone likes big, and it is not just big, but truly huge. If you think of the average insect as the size of a cat, *Ephemera danica* is the size of a cow. And it is this size that always stuns people the first time they see it; more often than not other insects' hatches are really hard to spot on the river surface as the insects are carried downstream on the current. But not the mayfly – it sits up on the surface like a yacht in full sail. Sometimes when out as a fishing guide I'll point my client in the direction of a solitary mayfly and they'll say, 'What, you mean that leaf?' Mayflies really do look like that at a distance, and I don't blame anyone for doubting their own eyes the first time they see one.

Your first mayfly day will be one of those defining moments in your evolution as a fly-fisherman or -woman. Cynical anglers dismiss mayfly-fishing as too easy, but anyone with an ounce of awe for a wonder of nature will be excited each and every year the calendar comes around to May, and that first time will always be hard-wired into your memory.

As with many things angling, the nomenclature is guaranteed to confuse, for the mayfly is far from being uniquely tied to May. On that grandfather of the chalkstreams, the River Test, the hatch runs from mid-May to early June. In a bad year it will last two weeks, in a good year over a month. However, 15 miles to the east, on the River Itchen, an almost identical river, the hatch will start in June and carry on well into July. Try an April walk through the water meadows alongside the River Avon and you will regularly see a hatch that lasts for just a day or so. Confused? I wouldn't blame you, but nature does not obey the calendar, and our mistake as anglers is to sometimes impose the parameters of human existence on insects that will live for barely twenty-four hours. As a mayfly you spend two years working up to that single day when you will hatch, mate, reproduce and die, so you had better get it right. In your nymphal state you don't care whether it is April, May or June, all you care about is the optimum conditions to do your bit to preserve your species on hatch day. You certainly don't have time to expand your horizons by travelling to another river, and therein lies the reason for the local variations. For in the same way that animal species have evolved on the different continents, our mayfly adapts to the tiny variations, such as water temperature or light intensity, within each river catchment. Unable to travel far enough to breed on another river within that single day, each successive generation reinforces the habits of those that have come before, preserving the unique characteristics of a Test, Itchen or Avon *Ephemera danica*.

But don't think our ephemeral insect is just a native of the chalkstreams. They occur all over the British Isles, and in Ireland they are the highlight of the famous dapping season of May and June. You will even see mayflies creating havoc in the riverside inns along the

River Thames as the after-work drinkers swat them away, fearing them to be some giant mosquito with a bite to match.

It is a racing certainty that your mayfly day will start with disappointment. Like any good angler fired up with enthusiasm for their first day on the river you will have read copious fishing magazines, spent heavily in tackle shops and boasted to anyone who might listen of the fantastic haul you fully expect to take. As the old saying goes, you will have 'risen early and created a mighty commotion in the household' and will arrive laden with kit and expectation. In your mind's eye you will turn the bend in the river to greet a steady stream of mayflies drifting down it, voracious trout greedily swallowing them down. You'll be thoughtful and measured, taking your time to rig the rod, tie on a new leader – the thin filament of nylon that connects the fly to the fly line – examine the hatch to precisely pick the right pattern from your box, and then with calm deliberation make that first cast of the day, so perfect that it catches your fish of choice the first time.

And the reality? Well, you probably won't wait until you get to the river before tackling up. In the car park you'll be all fingers and thumbs in your rush to get ready. New leader? Nah, you'll use the one left on the reel, though a little voice tells you this oversight will cost you dearly. Considered fly choice? Forget it. You'll put on the first fly that comes to hand or catches your eye. There is an old saying in the tackle trade that flies are tied to catch fishermen not fish. You have just been the first catch of the day.

The fact is that your 'mayfly madness' is already well advanced, but it's nothing that a couple of contemplative hours will not cure, for in your eagerness to get fishing you will have arrived at the river far, far too early in the morning to start. It is one of those things that coarse fishermen, legendary for pre-dawn starts, never quite understand about fly-fishermen: why we start so late. So late being around ten in the morning, which at the height of summer is a full four or five hours after sunrise, the time when any self-respecting carp basher or pike hunter will be contemplating getting his head down for a few hours' shut-eye. One of my first clients, Edward Bielby, an old-school city gent, liked to describe them as bankers' fishing hours: arrive at ten, go for lunch at twelve, be back a bit the worse for wear at two and pack it in at four. Clearly this was pre-Big Bang. Edward held to this theory regardless, which I think said more about his predilection for long lunches than the reality of the fishing conditions on any given visit. In the early days of our fishing forays I tried to persuade him to take a late lunch or hang around for the evening, but he never wavered. On the strike of noon, even in the midst of a purple patch when we were catching every cast, he would wind in, hand the rod to me and head off for lunch. Edward's view was clear: you set your own rules by which you fish. They might not suit others, but if they suit you, stick to them. The fish you don't catch will still be there after lunch or tomorrow. And if you are not fishing tomorrow then they will be there for the next lucky fellow. It is a lovely philosophy to live by, and though Edward was wrong to go home so early (evening fishing is often the most exciting of the day), he was certainly spot on for the morning start.

Yet the cadence of the fishing day is not determined by the angler, the fish or the weather, but by the insect life. There is no chicken or egg debate here. The insects start hatching, the fish start feeding and the angler starts catching. Understand this and you are on the path to successful fly-fishing. On a typical mayfly day everyone arrives too early. On this particular overcast, blustery mid-May morning my fishermen for the day, James and Olaf, and I peer down from the bridge to the fish below, who sit there glued to the bottom, gorged from their mayflies the previous day.

Logic tells you that surely a well-presented mayfly pattern, a careful imitation of what was on the river yesterday, will galvanize even the laziest trout to feed.

'What shall I use on them?' asks Olaf, who has applied that logic. My heart sinks, for I am certain that for the next half-hour every fly cast by Olaf and James will be ignored, despite numerous variations of fly, angle and presentation. But how can I tell them this? I can't – I don't want to dampen their enthusiasm – so I don't and we get into position.

From the outset Olaf struggles with a downstream wind and cannot make the heavy bulk of the mayfly turn over and land in a position where the fish can see it. A bit frustrated, he gives way to James, who tries his luck.

'What on earth am I doing wrong?' Olaf asks.

'When was the last time you picked up a fly rod?' I reply.

Olaf thinks about it for a moment. 'Last summer, I guess.'

'Tell me,' I say, 'do you play golf?'

'Yes.'

'Do you get on the practice range to hit a few balls before playing?'

'Yes.'

'Do you run?'

'More or less every day.'

'Well, I bet you do a few stretches first.'

'Of course,' says Olaf, 'but . . .' And then it dawns on him: fly-casting is like any other discipline. If you don't do it for a while you get rusty. You don't get bad, you just get out of the groove. 'I get it, you want me to practise!' I smile at him.

'It's not going to be a boot camp,' I promise, 'but let's give James some space and move a few yards downriver.' That's exactly what we did, and for fifteen minutes I took Olaf back to basics, casting with a bit of wool until he could land on a spot the size of a dinner plate at 15 yards. Then we tied the mayfly back on. A couple of casts later Olaf looked at me woefully as the old problem came back with the fly landing in the crumpled heap of tangled leader.

'Don't worry,' I said, taking the leader. I nipped off the fly, cut off a yard of leader and retied the fly. 'Try again.' This time all was perfect. The rod went up, the line stretched out behind Olaf, and when fully extended he brought it forward, and as the line and leader unfurled in a straight line the mayfly drifted onto the water.

'How did that happen?' asked a delighted Olaf, who repeated the same perfect cast to two or three different points on the river. The fact is, different days, different flies and different weather conditions demand different leader length. But when you've paid three or four pounds for a nine-foot factory-made leader it takes quite an act of faith to lop 2 or 3 feet off it the moment you take it out of the packet. Fortunately I hadn't paid for Olaf's leader, so I had no such scruples.

Back with James we guessed that the absence of any whoops of joy or splashing meant that he had had no success, but we abided by the angler's code that allows for admission of failure with a suitable excuse attached.

'Any luck?' asked Olaf.

'Nothing,' said James. 'They won't look at anything. I've tried at least half a dozen flies but . . . [the excuse is on its way] . . . they are clearly not feeding.' He reels in and asks me, 'What shall I try?' I am tempted to say coffee then launch into the explanation about why we need to wait until later in the morning, but they are so enthusiastic and fired up I don't have the heart.

'Well, it's clear the trout are not impressed by our mayflies and are waiting for the hatch later on as the day warms up,' I said, getting an explanation of sorts in, then pausing, hoping one or other might take the hint and suggest coffee, but they both just look dejected. 'So,' I continue, 'there is a little trick that sometimes gets them going.' This brings looks of hope and I produce a fly box.

Peering into the box I pick out a tiny little black midge, which looks exactly as you would imagine – one of those buzzy, bitey things you associate with the summer; they generally leave you with red spots. Actually midges are one of the few insect groups that hatch all year round so are a constant feature of the trout diet. Our trout know that the big meal of mayflies will be along later in the day, so for now they are just chilling out ahead of the feast. Yet like any diner offered peanuts at the bar or bread on the table the temptation to snack is irresistible.

James gives way to Olaf, who gets into position and after a few false casts easily covers the fish nearest to our bank. Not a twitch. My snack theory looks on shaky ground. Olaf looks at me by way of a question.

'Good cast, but leave him alone. Let's cover each of the fish in turn, from left to right.' (There were four spread across the river.) The second is equally dismissive, but the third twitches just a fraction. It is the tiniest of moves as the fish lifts a few degrees from the parallel as if to acknowledge the presence of the fly, but nothing more. If it wasn't for the gin-clear water and our absolute focus on that single fish it would be easy to miss, but we all three see it and let out a collective gasp.

Olaf casts again, this time the black midge landing a little further upstream, on the peripheral vision of the trout, which quick as lightning lifts off from the bottom. He drifts up on the current, letting the flow take him downstream, tracking the progress of the fly, getting ever closer beneath it. None of us breathe for we know that in the next moment the trout should open his mouth, suck in the fly and be on the line, but when his nose is almost touching the fly he has an abrupt change of heart and in an instant heads directly back to his spot on the riverbed.

'*Faen, faen, faen . . .*' exclaimed Olaf ('Damn, damn damn'), reverting to his native Norwegian, and then laid half a dozen increasingly frenetic casts over our fish. The trout was having none of it, resolutely refusing to budge until Olaf crashed down his line in a truly appalling cast, which sent all four fish fleeing for cover. But it didn't really matter, for just that one show of interest, even though the fish didn't actually take the fly, was a victory in itself and we were all content to take a break.

In the clear, alkaline waters of a chalkstream there is a hive of activity buzzing beneath the surface as a mayfly morning progresses. The thousands of little nymphs, which started life as eggs almost exactly two years ago to the day, are getting ready to emerge. Like a butterfly emerging from a chrysalis, a mayfly nymph stops feeding and prepares to head for the surface of the river. For our nymph this is the most dangerous moment in his life; as he struggles to shed his larval shell in the surface film, not only is he easy pickings for hungry trout, but get the timing wrong and there will be no other mayflies with which to mate – disaster at every level.

It must be said that part of mayfly madness is fuelled by just how obliging the nymphs are about their timing. For us anglers the delight of fly-fishing lies in the uncertainties, but just once a year to know exactly what will be hatching on any given day (with ravenous fish to match) is a special treat. Would we want it every day? No. Do we like it once a year? Most certainly. So, how on earth do the nymphs arrive bang on time every year?

You'll hear all sorts of theories why – light intensity, UV rays and water temperature – and I guess there is an element of fact in them all, but more realistically do we care that much? The important fact is that from around the middle of May is the time to head for a river with a box of mayflies and great expectations.

By now it was well past eleven o'clock and whatever expectations Olaf and James may have had were morphing into disappointment as the high of the fish-we-didn't-quite-catch faded. Sometimes as a fishing guide this is a worrying moment, and if you are not careful you can lose your clients to surly indifference for the rest of the trip. On any other day I might have been worried and have tried to convince them to get back on the river, but not today. Today I'd have staked my life on the certainty that within the next hour or two an armada of mayflies would be floating down the river to give Olaf and James the fishing day of their life. So I fuss around, fill them with coffee, and we compare fly boxes to fritter away the time while I constantly scan the river with one eye.

Sure enough, within half an hour the first mayfly of the day appears; it is strange, but they do just 'appear' like a magic trick of nature. One moment you have a flat, dull river surface and the next the iconic vision of what fly-fishing is all about. The morning sun bounces off the dappled white wings, making it easy to track the progress of the mayflies floating downstream.

'Is that a mayfly?' asks Olaf. 'I can't believe how huge it is.' James picks out a mayfly from the box and holds it up at arm's length to compare the silhouette with that of the real mayfly.

'I thought these flies were a joke, but they really *do* look like the real thing,' he says, and wastes no time tying it onto the end of his line.

The three of us stand watching the stately progress of the solitary mayfly as it drifts further downriver. It seems inconceivable that such a tasty morsel will be left unmolested for long, and sure enough after 15 yards, with a gulp and a splash, the best meal of the day disappears into the stomach of a trout.

'That's it, no more coffee and sitting around,' I announce, so we gather up our stuff at speed and head down the back path of the woods that will bring us out to the bottom of the beat. By the time we are in position the hatch is beginning to build, with a good ten or a dozen mayflies on the surface within casting range of James. Rather than cast blind we wait for the first to be taken by one of the fish we can see holding in the current. Sure enough we see a brown shape loom up from below, open its mouth, suck in the fly, turn its head down and with a slash of his tail drive back down into the depths, using the force of the water to push the mayfly down its throat. It is all over in two seconds, with the same trout back on station and ready to take the next fly that comes along. But this time it will be James's fly.

The first few casts are all over the place, too far left, too far right or not far enough ahead of the target fish. I think James is being overcautious; a mayfly is a big offering for any trout and it needs to eye it up before committing, so it pays to land the mayfly a good couple of yards ahead of the fish. With a little encouragement this is exactly what he does, and we see an action replay as the trout locks on to James's fly. But this time as he turns his head down to swallow the fly Olaf and I in chorus scream 'strike'. James's rod comes up and the fish is on.

The fight is short and James soon has the fish in the net. He holds him up for a quick photo; after all your first fish on a mayfly is a must for every family album. Back in the water we watch the trout shudder as he coughs up three or four mayflies in various stages of digestion before heading for a clump of weed. He will no doubt sulk there for a few hours before getting back on the hatch.

Next up is Olaf, and while James emails the fish photo to anyone who might be remotely interested, and probably plenty who aren't, we wait for the next rise. Sure enough we don't wait long, and Olaf needs no invitation. Learning from James, his first cast is bang on and a few minutes later we are releasing his first fish.

At this point in the day my job is done, for a while at least, so I sit down on a fallen tree to watch the action. The boys fish turn and turn about, each giving way to the other once a fish is caught. Not every cast catches a fish and not every fish wants to rise, but that's part of the fun. Find a fish that won't react to your fly? Well, tear it off and try a different variation. Have a fish that comes up and looks but will not take? Next time give it a twitch. A fish that simply won't play ball? Move on to another; there are plenty more fish in the sea (well, there are!). The fact is that part of mayfly madness is that you have more choices for more fish than at just about any other time in the fishing year, so it's a chance to be a bit blasé or to try some new tricks.

Of the two, it is James who tires of catching first. Tired of catching fish? Yes, it really can happen, and he joins me on the tree to watch Olaf, who is soon into another fish. As we watch it splash around, James asks, 'Simon, why is it that despite the thrashing around and all the fish we have taken out of this one spot the others don't seem to care?'

It is a bizarre thing and he's absolutely right. If I lobbed a tiny stone into the river all the fish would flee in an instant, but to have one of their fellows tear up and down on the end of a line and then disappear doesn't seem to bother them one jot. There is clearly no empathy between fish, and long may it continue from an angling perspective, but as for an explanation, I don't have one.

As is often the case on a mayfly day, after the big surge the hatch eases off to the point where the flow of mayflies becomes a trickle and the trout lose interest. But it is simply a lull in the action; there is much more to come in the afternoon, so we head back for lunch. James and Olaf chatter away ahead of me, comparing notes from their successful morning. I know for sure it is a couple of hours they will remember for ever. I'm happy for them and feel just a little surge of pride at being able to lift the curtain on something most people never get to see or be a part of.

'Tell me something,' James asks, 'we've seen hundreds, no probably thousands of mayflies this morning, only a tiny proportion of which have been eaten by the trout and ducks [yes, even ducks love to eat mayflies], so where have they all gone? As far as I can tell they have vanished.'

'Well, not exactly,' I say, stopping beside an alder bush and lifting a branch. 'Look here.' And there, as I expected, hanging upside down on the underside of the leaves are two or three mayflies on each leaf. They are completely unfussed by our intrusion and make no attempt to fly away. I gently pick one up by the tips of its wings and place it in the outstretched palm of my hand so that

James and Olaf can examine it. The mayfly sits up on my hand, crouched on olive-coloured legs that resemble the hindquarters of a cheetah about to spring. Its three grey-black tails are double the length of its body and its sail-like cream wings, translucent with heavy venation, sit on a creamy yellow body that is long, thin and maggot-like. Look directly at the head and huge bulbous eyes stare back at you above a jaw that would easily crush anything unfortunate enough to get in its way.

'Creepy,' says Olaf. 'Any idea what they taste like? Trout clearly like them.'

'I have heard it said they taste like butter, but I've never quite had the balls to give it a try. How about you?' I push my hand towards Olaf's face, who backs away laughing. 'I would of course, but I'm a vegetarian.' I don't think he is really, so we carefully return our mayfly to under the leaf, where he will remain for an hour or two before shedding his skin and emerging looking much the same but ready for the all-important mating ritual and the famous mayfly dance.

The odd thing about the mating dance is that it starts all of a sudden. One moment you are looking in a particular direction, maybe at the treeline or over the water meadows; turn away then look back a few seconds later and the air will be alive with thousands, possibly millions of dancing mayflies. The air can get so thick with them that it is like gunsmoke drifting across a battlefield. But this is no random dance; its beauty lies in the precision of nature's pageant.

This is the cue for James, Olaf and I to get ready for the really serious fishing of the day. As the late afternoon sets in, the mating is complete and the males slink off to die in the bushes. The females will soon return to the water to lay their eggs, where the trout will be ready for the most almighty feast as the mayflies spend the last few minutes of their lives laying down the seed for the next generation.

I make sure that both James and Olaf have two particular types of mayfly for this final hour or two; the first looks a lot like the morning flies, to imitate the female sitting on the surface, albeit this time she is laying eggs rather than having just hatched. The other fly is a very plain affair, looking like one that has been pressed in a book. It represents the dead mayfly lying in the surface film, all stretched out like a corpse floating in a Hollywood 'B' movie swimming pool.

I am pretty sure that my guiding role is over for the day; my anglers have cracked the mayfly code and the fish will be certain to oblige. It now comes down to waiting for the fish to feed, banging the right type of fly down in the right place, then changing between the two patterns as the fish switch from the live to the dead insect as their fancy takes them. So I position the two on different parts of the beat and wait at the mid-point for the action to begin.

We don't have to wait long for the first egg-layers to head for the river. But stuffed full of over 8,000 eggs, they are no longer the nimble fliers we saw at the dance.

They adopt one of two strategies to deposit the eggs in the river; one keeping them alive a good deal longer than the other. The suicide strategy is to sit on the surface to lay their eggs, but the temptation to the trout is too great and there will be no chance of escape as the eager trout hoover them up or the mallards line up across the width of the river to do the same. The competition is sometimes so frenetic that a trout will snatch a mayfly from beneath the beak of the duck, who will cock his head with a mystified air as if to say, 'Who stole my mayfly?'

The savvier mayfly descends to the surface of the water to release her eggs in groups by dipping the tip of her abdomen onto the surface at intervals, using the surface tension of the water to draw the eggs from her body. But even this is no guarantee of survival, for the trout will leap from the water to grab the mayflies in midair. Whichever path our female chooses, her lifespan is now measured in minutes rather than hours, and as the evening draws in the river surface becomes thick with mayfly corpses.

While James seems to have struck a purple patch, catching a fish with almost every cast I can see, Olaf is getting more frustrated as the fish continue to rise in front of him but ignore his every offering. This spills over into his casting, which gets wilder with each cast, his line swishing like the tail of an angry cat. So when his back cast gets caught in the bushes behind him for the third time, this seems the moment to go to his aid.

Using the excuse of stripping down his leader and putting on a new tippet is my way to calm things down, so while I clip and knot on a new fly we stand on the bank examining the water surface picking out the egg-laying, dying or dead females in turn.

'What's that?' asks Olaf, pointing to a perfect circle of pulsating ripples, about the diameter of a coffee cup.

'That's a dying mayfly – in its final convulsions as it drowns,' I say. 'Watch it.'

The violent shudders last a couple of seconds, the mayfly enduring them two or three times – as if absorbing a giant electric shock – until it lies still; dead on the surface. As if on cue, a trout rises from below in the most nonchalant manner, knowing that the mayfly is a spent force with no chance of escape, and opens his mouth to let the flow of water draw the meal into his jaws.

'Let's try this,' I say to Olaf, holding up a spent mayfly pattern. 'Pick any fish you like!'

There is indeed a lot of choice: half a dozen fish rising at regular intervals within 15 yards' casting distance. Some were feeding on the egg-layers, others waiting for the pulsating mayflies, while the rest lazily sucked down the dead. Trying his luck with one of the middle group, Olaf casts twice over a feeding fish that ignores the fly.

'Next cast, when your fly gets into the view of the trout, give it a twitch.' Olaf looks at me as if not quite understanding. Maybe it's just a language thing, but not knowing the Norwegian for twitch I try a different tack.

'Waggle the tip of the rod to make the fly move as it drifts.' It takes Olaf a couple of goes to get it right, but boom, the moment it happens the trout comes straight up without hesitation and takes the bait.

'Ha, ha, ha,' screams Olaf with delight. 'Amazing, so simple and so damn effective.' And for the next half-hour we have some fun picking out a particular fish feeding in a particular way and changing our strategy to match, but the twitch has become Olaf's secret weapon of choice.

James walks back down the river to join us with a huge smirk on his face, holding out a bedraggled fly.

'What's that?' asks Olaf.

'Well it was a Thomas's Mayfly, but after ten casts and six fish it is now my favourite fly *ever*!'

Not wishing to be outdone, Olaf says, 'Watch this', and goes on to demonstrate his now perfected 'twitch' to deadly effect. As he gently slips his umpteenth fish of the day back into the river, he turns to James. 'Want to have a go?'

'No, I think I'm done. And this fly,' he says, snipping it off and sticking it in his cap, 'is going to be my lucky charm.' And like gamblers who have been on a hot streak, walking away from the table with pockets bulging with chips, the pair decide to call it a day.

10

CRAYFISH INVASION

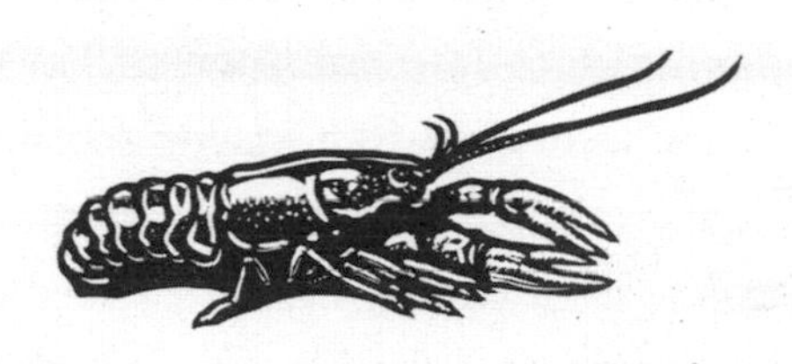

NOT EVERYONE AGREES, but June gets my vote as the best month of the year. River keepers dread it because it is the heaviest and hardest weed cut of the year. Two weeks of hell most will tell you, and for once they are not exaggerating. Fishermen dislike it because a post-mayfly torpor seems to grip the fish. Suddenly after a month of savage feeding and in-your-face fly hatches the fish get hard to catch. People will tell you that the trout are stuffed full of mayfly, unable and unwilling to eat anything for many weeks to come. This is patently untrue. Try it yourself sometime. Gorge yourself to a standstill and then see how many days you can go without eating. Not more than a few I would guess. Trout are no different to us in this respect. The truth is that the fish go back to type for the other eleven months of the year: cautious and selective. As fishermen we have forgotten this, grown a little complacent and lazy in our habits. We need to reset.

But for me June, aside from any fishing niceties, is the month when the chalkstream valleys are at their most sublime. At every turn the meadows are wrapped in colour. This is still early summer. Everything is bright and lush, that dustiness of late summer still a while away. The rivers shine and glint like sterling silver. The days get longer as the evenings get later. It hurts me to leave the river whilst there is still light in the day, so I usually don't, choosing to stumble around in the dark to find my way home. It makes for a long day, especially when half of the June days are spent wielding a scythe and manhandling rafts of heavy weed, but it is really no hardship. The work must be done, and against the backdrop of such perfect countryside how bad can it be?

Weed-cutting is one of those oddities that as an organized and vital part of river management are pretty well unique to the chalkstreams. At fixed periods during the season when the rivers close to fishing, we don our waders, pick up the hand scythes, step into the water and for as long as two weeks at a stretch cut, trim and clear the weed that grows up from the riverbed. Often when I mention it to people unfamiliar with the term, I can see the cogs in their brain whirring to compute the concept Weed? Cut? River? Really . . . They are right to be baffled. After all, where else in the world would the river owners devise a system whereby out of a seven-month fishing season, one and a half months will be out of commission?

The success of this unusual arrangement is part historical, part legal, part necessity, and with a bit of peer-group pressure thrown in for good measure. The historical has nothing to do with fishing. The practice of neighbours cooperating to coordinate cutting goes back centuries, to when there was a watermill every couple of miles on every chalkstream. Cutting the weed that impeded the passage of water down the river was a necessity for the millers, but if you failed to tell your downstream neighbours of the cut weed heading their way it would very soon jam up the waterwheels, so to avoid angry disputes a progressive rota from the headwaters down to the estuary came into being. As it turned out this suited the fishermen just as well as it did the millers, so when the milling died out the organized cutting rota remained in place. Whilst this is simply common sense, creating a practice that spilt over into fishery management, the legal is more esoteric.

In most nations around the world rivers are public property; a free-for-all where nobody owns the fishing rights and the fish that swim in the river are there for anyone who wants to catch them, usually only for the cost of a government fishing licence. Sometimes the access to fish the rivers is nuanced by ownership of the land that borders it, with ownership rights conferred on someone down to the high-water mark. But generally speaking, if you can get onto the river from a road or some such it is yours to enjoy, be you an angler, swimmer, canoeist or whatever takes your fancy. But on the chalkstreams, and most other rivers in England and Wales for that matter, quite the reverse is true. The rivers, access to them and the right to fish them is in private ownership. Every inch of bank and every square foot of fishing rights from the moment the water emerges from the ground to the tidal estuary is owned by someone, enshrined in ancient statute, and if you are one of those lucky few you have the right to call yourself a riparian owner. Is this good or bad? You can argue the case on many levels, but the fact is that by owning a river you have a vested interest in looking after it. Aside from the obvious point that they are valuable (good beats change hands for millions), it is hard to own a river and not want the best for it. And you don't have to be a riparian owner who fishes for this to apply. A badly tended river tugs at the heartstrings as much as a once beautiful but now neglected garden.

On a more practical level the chalkstreams have a vital role as irrigators of the river valley, and cutting the weed regulates the flow. Leave it to grow unchecked in the summer and gradually the river will become so clogged that the banks will burst and flood the water meadows, either destroying the hay crop or denying grazing for the cattle. If you are both the riparian and landowner, aside from any fishing interest, the cutting becomes as much a part of the farming routine as hedge-cutting or harvest. If you own just the river you could of course ignore any such obligations, but the ideal of being a good neighbour is still alive and well in rural Britain.

Weed-cutting at Gavelwood starts on the second Monday in June; it has been thus for as long as anyone can remember, coinciding with the end of the mayfly hatch and Duffers Fortnight. In a way it is something of a relief for us who work on the river, despite the hard, physical labour ahead. For the past six weeks it has been relentless. Fishermen every day. Early starts to get everything ready before they arrive, late finishes to see them home. It is not bad in any respect – sharing the excitement of some of the best and most unique fishing on the planet is a joy in itself – but I know the selfish parts of us all yearn to take the river back for ourselves. As get it back we do, in spades.

Essentially there are four ways to cut a river; chain scythe, hand scythe, pole scythe or weed boat. The last is regarded by most as the least worst option when the first three are not possible. There are parts of the Evitt when you get close to the sea where it is too wide to work from the bank and too deep to wade, so the boat is the only option. The flat-bottomed boat is driven by paddle wheels on either side operated by a man who sits in the middle raising and lowering a cutting bar that looks a bit like a giant hedge trimmer, which is fixed to a hydraulic arm in front of the bow. It is a mighty efficient way to cut weed; a good operator can do in a few hours what would otherwise take days back in the era when the only other option was hanging over the side of a rowing boat with a scythe. But that said, it is something of a blunt instrument and sometimes ecologically destructive; chugging along with limited manoeuvrability, it is difficult to discriminate between the weed you want to keep and that you wish to cut. Subtle changes in the riverbed, so easy to see when wading, are harder to fathom. The other option, chain scythes, have many of the same drawbacks, plus they are damned hard work. Standing on one bank, with your partner on the other, dragging two hundred pounds of submerged steel blades upstream with a sawing motion for a day is as much hard graft as most people do in a year. If you think the blades cut like the metaphorical hot knife through butter, think again. It is more tearing than cutting as they drag along the riverbed, every so often becoming so entwined with weed that you have to drag the whole kit and caboodle onto the bank to clear it before hauling it back across the river to start again.

I don't much care for the word 'cut' in relation to the work we do to control the river weed; it is the one we always use, but it smacks of destruction or removing something bad, like cutting out a cancer. That venerable river keeper on the Wiltshire Avon Frank Sawyer had it right when he referred to the cut as mowing. It sounds so much more positive, more grooming than cutting. Taking something that is already good and making it better. In fact the hand scythes we use today used to be called 'mowing tackle', the very same ones that one man and his dog Spot of nursery rhyme fame went to mow a meadow with. So it is with these in hand that we all gather on thc first morning of the Gavelwood June cut at the Drowners House.

There is never any great rush to get going; we all sort of need to decompress after the mayfly. In truth we haven't had much time to be together or gossip over the past few weeks, so this is a time to catch up. Inevitably the conversation is a mayfly post-mortem. Was it early? Was it late? An average year, or better or worse than most? That everyone has an opinion is what makes it fun, plus memories of seasons past and the comparisons we make are never perfect, so there is plenty of scope for disagreement. Eventually we will coalesce around a consensus, and that will be another mayfly consigned to history. If you work on a river you have to be a bit pragmatic in that respect. What is gone is gone, and the anglers who come next week will not thank you for telling them they should have been hcrc last week.

As the others took whetstones to the scythe blades to get them razor-sharp I walked the length of Gavelwood to assess the 'mowing' we needed to do this year. No two seasons are identical, and over a period of a decade you can have anything from years when the June cut is almost a non-event to others when the river is clogged with weed, the two-week allotted period barely enough time to complete the job. It is the sun and water that dictate how much the weed grows from late April to early June, during which time all cutting is embargoed. Plenty of sun plus well-oxygenated water in the spring, and boom, the surface of your river will be a sea of little white flowers where the crowfoot has grown like Topsy and broken out onto the surface. The fish just love these years; plenty of cover and plenty of food all concentrated into the fast rivulets between the weed beds. As you might imagine, this is a hard one for the keepers. On the other hand a dry winter followed by a dry, overcast spring and large sections of the river will be devoid of weed, with at best meagre growth covering only sections of the riverbed. Fish hate these years, fleeing to the deeper sections for cover and food, whilst the keepers make the best of a bad job, preserving the weed they have the best they can and hoping the conditions improve.

For the angler the sparse years make the fishing easy, but the catching harder. An open river with no weed to get caught on when casting, or heavy beds that a hooked fish might bury itself in, might seem attractive at first glance, but without the cover and food that the weed provides the few fish that remain will be easily spooked. On the other hand casting into those narrow fast runs between weed that sits on the surface requires great accuracy. Err a few inches to the left or right and you get snagged. Hook a fish on the wrong side of the weed and he will be a devil to land. Hiding in the dark shadows beneath the weed beds the fish are hard to spot. All in all, when it comes to nurturing and cutting the weed the river keeper has the unenviable task of finding a way between these two extremes. This year at Gavelwood we are at neither extreme; it is going to be one of those years where we have to cut to open up the river for the fishermen whilst at the same time preserving enough weed for cover and to hold up the water levels.

After weeks of seeing the river from the bank, it suddenly looks very different from in the water. More beautiful. More serene. More alive. The crowfoot flexes in the current, gently pulsating, fractionally raising and lowering the height of the water, the little fronds of the weed skittering in the wake of the small waves the movement creates. Dozens of tiny olive insects drift down on the water, spinning on the surface as they get caught in the fast, irregular flow of the rivulets between the weed. I cup my hand to scoop one up who shows no sign of concern or indication that he wants to take flight just yet as I hold him up to the sun. His arched body is rigid, the three long tails keeping him poised and upright as the large pale wings catch the gentle breeze. Along the banks the tall yellow flag irises are the plant of the moment, but the hairy willowherb and water forget-me-not that tumble over the bank are the flowers that catch the eye from river level. And all the while from behind the barbed-wire fence the cattle stare down at us, chewing the cud as if ruminating as to why we are in the river and not them.

The first slice of the scythe through the water and a long tail of crowfoot is a satisfying moment. The freshly honed blade truly does cut like a hot knife through butter, exposing the white ends of the cut weed that hangs immobile for a fraction of a second until the current catches the cut weed, the buoyant stalks bob to the surface and the detritus starts its long journey to the sea. From time to time the cuttings will get caught, their downstream progress halted until they free themselves again or another river keeper sends them on their way with a shove from a weed rake. And so a new chapter opens for the myriad bugs that live in the cut portion of the weed, the cutting displacing them from their place of birth. Surprisingly they tend not to cling on to the safety of the floating weed for long, preferring to swim away at the earliest opportunity to find a new static home, and in a few hours most cut weed will be devoid of any invertebrate life. But in these few hours a vital exchange will have occurred. The deeper water where the weed doesn't grow will be seeded by the displaced nymphs who depart the floating weed, and the general mixing up of the populations reduces the chances of interbreeding, which brings in new blood-lines to improve the gene pool.

At Gavelwood we cut in pairs, working two abreast across the width of the river cutting in tandem. The idea is that we are creating the perfect home for a trout; the weed is where he can hang out when resting or hiding from predators, whilst the open areas are ideal when he is on the fin, looking out for food. We are cutting the weed for the fisherman as well; in an ideal world we'll want him to be able to cast his fly so it lands over the just submerged weed bed to then drift into the open area where the fish is looking up for food. So as far as we can we trim the weed so that it doesn't break the surface. Beneath the surface the zigzag path of the water created by the pattern of the cut suits the nymphs very well. The fast channels between the weed beds cleanse the gravel, keeping it well oxygenated for the eggs that are incubating. And once they have hatched they can hide between the loose stones until they are big enough to head for the weed cover. Sometimes we will break with the chequerboard pattern, which hold back the flows and create some depth in areas that might otherwise be too shallow in low-water years.

The cutting is the easy and satisfying part of the job; the clearing down is just plain hard, but no less important. Inevitably the cut weed gets caught up on everything you can imagine: uncut weed, bridge supports, branches that hang into the water, shallow gravel sections, tree roots, back eddies . . . we try to make our lives easy by starting downstream and working upstream to make the cut, but this only helps so far. So by the middle of the afternoon we will lay down our scythes and pick up the long-handled grab rakes, progressively loosening the caught weed and sending it on downstream. And there is the rub really. Not only are we moving on the weed we have cut, but also the weed that has been sent down by the keepers cutting above us. And woe betide any keeper who fails to clear down each day, instead waiting until the end of the week to send down huge and unwelcome rafts of weed to his unsuspecting neighbours downriver.

Oddly enough the weed-cutting activity doesn't seem to bother the fish that much. On days of good hatches they will continue to rise until you could almost bop them on the nose with the tip of your scythe, and it amazes me that they will return to their usual haunt almost as soon as we have waded past. Fish are strange creatures sometimes; one day a single splashy cast will spook them for hours, but wade through their home, slice away half the weed they live under, and hey, no problem, I'll be back in a minute or two, they seem to be saying. In truth the activity we stir up with the work, both the wading that disturbs the gravel bed and the cutting itself, creates a food bonanza for the trout. Unlike grayling, who are foragers, trout want service. The lady of the stream is quite happy to poke her snout into the riverbed to dislodge food; not so brown trout, which prefer to wait for the food to come to them. So all our clumping about is heaven-sent for the trout who will take advantage of the momentarily homeless nymphs by cruising under the cut weed to take advantage of this seasonal windfall.

Our general activity also disturbs the huge and ever-present population of freshwater shrimps, which are, perhaps a little surprisingly, one of the staple parts of the trout diet. Like flamingos that need shrimp to pigment their feathers pink, so it is with trout, whose flesh is a pale pink from the shrimps and other crustaceans they devour. It seems contradictory to talk so readily of crustaceans as an everyday part of life in a river in the context of the chalkstreams – it is a term you might more readily associate with the sea – but they are as much part of everyday life as the nymphs and fish. Look in a fisherman's box for a clue. Amongst the delicate dry flies and practical nymph imitations you will see something that looks just like a shrimp, or if you prefer a tiny prawn, which is about one quarter the size of your small fingernail. It is tied on a curved hook, and with a plastic body that shines when wet it looks every bit as real as its bigger cousin you'd see on the fishmonger's counter. It is a fly (we call it that although it so patently isn't) that anglers use all year long, sunk deep in the water to imitate the scavenging shrimp who likes to forage close to the riverbed when he is not hiding out in the weed or gravel.

Even something as apparently indigestible and unpalatable as a snail will catch the attention of the fish from time to time. Aquatic snails need calcium-rich water, so the chalkstreams are heaven-sent in this respect, and as they need highly oxygenated water to breathe in the attraction becomes greater still. In normal times at Gavelwood we will not see much of the snails; they pop up to the surface to fill their lungs, then sink down to feed on the weed. But they have a weakness that makes them vulnerable to trout. The ecosystem of the snail is a hopeless converter of air from water, and at certain times in the summer, when the water temperature rises by a few fractions of a degree the oxygen content will drop to the point that the snails must abandon the depths to hang in the surface film by their feet. Stuck suspended, especially in the shallows and back eddies in August and September, they are low-hanging fruit for the trout, who hoover them up in terrific numbers, swimming with their bodies half out of the water as they patrol with open mouths. Catch a trout on days like this (yes that same fly box will have a snail imitation) and the belly of the trout will feel like a well-stuffed bean bag.

If you thought an imitation shrimp or snail pattern sits uneasily in the lexicon of a fly-fisherman I'd hazard that a fake crayfish makes it more confusing still. Who'd think that these mini-lobsters have any part to play in the daily life of an English chalkstream? But they do, and the population that thrives today is an exemplary case of what might go wrong, will go wrong. Strangely the story starts not in North America but Sweden in the 1960s, where the population of the native Scandinavian noble crayfish had been declining for the first half of the century. For the British any decline of our own white-clawed crayfish at around the same time might well have gone unnoticed, but among the Swedes this arthropod was harvested in its millions, very much part of the cultural and economic life of the rural regions. So in 1969, to fill the gap, the government took delivery of shipments of North American signal crayfish. As a plan it worked great; the newcomers thrived in the Swedish lakes and rivers, growing faster and bigger than the native nobles. What nobody knew was that the imports carried the crayfish plague *Aphanomyces astaci*, a water mould that wiped out the natives wherever the introductions took place. However before this was known the signal had gained a reputation for being a fast and easy grower, so the introductions spread across the entire country. It took a full fifteen years for the facts to sink in to the point where imports stopped in 1994, and by that time the damage was done, the nobles all but extinct.

Across the North Sea, soon after the Swedish imports began, word spread of this wonder crop, and it was not long (1976 to be precise) before the first signal crayfish arrived in the UK, not to make up for any decline in our own population but as a get-rich-quick scheme that was well advertised in the press. The pitch was essentially this: if you own a lake or river, seed it with a few North American signal crayfish, leave them alone to breed for a couple of years, then reap a lucrative harvest for many years to come at almost zero cost. Too good to believe? Well, the proposition was correct in many respects. Crayfish do breed easily and fast. Once they are in a body of water nothing stops them or eradicates them. Catching them is easy. Lob out a lobster-style pot overnight, bait it with just about anything (a partially opened tin of cat food works well) and you will have a full pot the following morning. So far so good. The problem was that for a get-rich-quick scheme you need buyers, and back then (this was long before the days of Pret A Manger's crayfish tail and mayonnaise best-selling sandwich) there really wasn't a mass market or appetite for crayfish, so the fad was short-lived – unlike its effects on the native white-claw. Britain is a small country, where the waterways, canals, rivers and lakes are either interconnected or separated by relatively short distances, so once introduced the North American signals spread like wild fire and the same plague that did for the Swedish nobles is now wiping out the native British white-claws.

At Gavelwood we don't have native crayfish any longer – I am told they disappeared decades ago – but creeping around on the riverbed we have plenty of signals. I guess I should be rather angry at them for taking over, that 'non-native' tag a badge of shame, but it's not really their fault and on the Evitt at least, they don't seem to do any great harm, though I am wary that if the population gets out of hand the salmon, trout and bullhead numbers might suffer. True, one of their favourite foods is caddis nymphs, but the decline of the sedge population is equally great on the chalkstreams without crayfish, so it would be unfair to lay that at their door. On the canals I am told they weaken the banks with their tunnelling, and coarse fishermen curse them roundly as they strip their hooks of corned beef, sweet corn and maggots, but here at Gavelwood I find them endlessly fascinating.

They are not the easiest of creatures to spot; with that dark brown shell they are well camouflaged, and tending to hide in dark recesses and move by stealth they mostly go unobserved. Sometimes greed gets the better of them – fish eggs always get them going – but they are most of all the vultures of the riverbed. They don't exactly circle the dying, but corpses, be they fish, fowl or animal, get torn apart by their large front claws. For something that only scuttles along on the riverbed they have a remarkable propensity to spot humans. How they do it I have no idea, but they can move pretty quickly when they see me. That said, picking them up before they get away is easy as long as you time it just right, clamping them with thumb and forefinger where the body ends and the tail begins. What I love about them is that they get so angry when picked up. They have no sense that you might be bigger, smarter or stronger than them. They twist between your fingers like an angry steel cable, frantically waving their eight legs about and fixing you with the purple dots of their eyes. They try to arch back to grab you with their pincers. If they succeed it actually does hurt, and I for one give up the struggle at that point.

The females are usually more docile than the males; turn one over during the winter and early spring and you will see the eggs in clusters under the tail, very much like a lobster. And then a month later do the same to see the extraordinary sight of the hatched eggs transformed into tiny infant crayfish that stay attached to the tail until June. Once free of the mother the troubles for the young are legion. Everyone likes a baby crayfish, which looks like its parent, just much, much smaller and translucent white, like a living crayfish skeleton. Naturally enough fish eat them when they grow a bit older, but in the infant state carnivorous nymphs like the damselfly have a great liking for them. Things don't get much better as the crayfish get larger and grow a hard shell; the beak of the heron spikes them easily with one jab and otters will eat them all day long whatever the size. In fact against these odds you have to wonder how they survive at all, but that is one of the great things about a chalkstream – there are so many niches that all sorts, however unexpected, find a place in the hierarchy.

It is strange, but the whole river seems to take a collective pause for breath around mid-June. For us who work there we will have broken the back of the weed cut and we can throttle back for a few days whilst we trim up the river, send down the last of the cut weed, and still have the place to ourselves before we reopen for fishing in the third week. The fish are definitely less frenetic – the memory of the gorging weeks of the mayfly hatch soon fades, and with nearly eighteen hours of daylight and with the living larder of the river writ large – nymphs, shrimps, snails, beetles and insects at every stage in the life cycle abound – they hunt for food when it suits them. The creatures, birds and animals, are more at ease with their world. The water voles are by now well into their second litter and maybe a third. The imperative to perpetuate the breed is sated; admittedly perhaps only one third of those born have survived to this point in the year, but in the numbers game that they play to repopulate each year those odds are plenty good enough. The wild fowl are done with nesting, except perhaps for the occasional duck or moorhen that rather mournfully sits on a nest, the first attempt having been stripped of eggs by a greedy stoat or some such river raider. But for those who avoided such a fate, life is very much easier; the fledglings are halfway to adulthood, beyond any real help that might be provided by their parents, and it seems they stick together more out of habit than need. The exception to these rules is most definitely the swans, and for a bird that has nothing to fear from anyone it makes you wonder why.

Maybe it is because swans mature much more slowly. They will not breed until their third year, whereas most other birds do so in the second year or earlier. It is certainly nothing to do with size. Even though by June the cygnets are still half the size of their parents they are bigger than just about every bird on the river bar geese, who are their only competitors. It is extraordinary how tight the swan family group remains right through until the autumn. At Gavelwood we have our resident pair that hatch anywhere from four to eight cygnets; inevitably a few fall by the wayside from predators like foxes, who will grab them when they roost on the bank at night, or disease. Wet, cold summers seem to be bad in this respect. This year we are at three, and by this point the really dangerous months are behind them, but they stick like glue to the parents. At any given time, day or night, there will never be more than a few yards separating the five. If anyone or anything comes in range they start this strange sort of sotto bark-cum-cluck that they fire at each other constantly until the perceived danger is past.

Perceived is definitely the word; the only creatures that go near swans are geese, and the two are most definitely warring tribes. Geese are really the only birds that compete with swans for food. Both of them like to graze the riverbed for the weed, snails and insects, though with a neck twice as long, the swan will always outcompete the goose for the choicest, lushest shoots. Maybe it is some need to prove themselves, but during the summer the geese seem to go out of their way to infuriate the swans. They only do it in groups – I guess safety in numbers – but you can see the tableau evolve over a morning. The fact is that the Evitt has more than enough grazing for both groups, but working on the principle of the grass always being greener the geese will gravitate ever nearer to the swans, who will gather tighter together until one goose will, without a doubt deliberately, stray into their midst. The cob will not tolerate this for more than a few seconds before chasing the goose away, and at this point all hell lets loose as the other geese retaliate by charging at the swans. It is probably the bird equivalent of a pub brawl that starts with the words 'Oi, you pushed my mate'. The river explodes into a brawl of thrashing wings that beat the water white as everyone lunges at everyone else, spray misting the scene. This is definitely sparring rather than fighting; the two types of bird rarely make physical contact beyond clashing wings. Only once have I seen a true fight, when the swans cornered a young goose, repeatedly pushing it under until it either drowned or died of fright, which I could not tell. But the clash is always short-lived and the outcome the same. The swans will regroup at exactly the same point they began, waggling their tails in a feathered V sign at the geese, who quickly retreat, sloping off to a safe distance chastened, unharmed but determined to do it all again. What was the point of all this, who can tell? The one thing I do know for sure is that the lives of all of us on the river – people and creatures – are dictated by the turning of the seasons. Sudden changes, subtle changes. They all matter, and Midsummer's Day, the summer solstice, is a day I approach with mixed feelings. It should be a celebration of everything that is great about the chalkstream summer. Long, warm days when everyone harvests the sun and the easy living it brings, but somewhere at the back of my mind there is a little voice that reminds me this is the high point of the year. There is more that has gone before than lies ahead. The days get shorter rather than longer. Animals and plants have done most of their growing, from here on in it is about consolidation. Maybe I am being a misery. Certainly in angling terms I have no reason to be gloomy. Looked at sensibly, the mayfly apart, we still have three and a half of the most glorious months ahead compared with the one and a half just gone.

I suspect I am not the first to have these feelings. Not far from the banks of the River Evitt is Stonehenge, where modern-day man makes a huge fuss of the summer solstice. That is all fine and all great fun, but the truth is that there are plenty who hold to the belief that the circle of stones was not built to celebrate the height of summer but the depth of winter. On reflection it makes a certain amount of sense; for Stone Age man five millennia ago the winter solstice was far more of a turning point in the year, the moment when the first half of the winter was over and the march towards spring seemed just a little bit shorter from every day thereafter. But for all my glass-half-full attitude to Midsummer's Day, the shortest night is the one we celebrate to mark the annual return of that prodigal, the sea trout.

11
MIDSUMMER'S NIGHT

SEA TROUT GO by all sorts of names. If you live in Wales it is sewin. In the West Country peal. In Scotland finnock. Down the northeast coast of England whitling. Add into the mix salmon-trout, herling, school peal, harvesters and slob trout (there are more names, but enough for now), and you might be forgiven for wondering how one fish can have so many names. After all salmon are always salmon, brown trout are always brown trout, so what is it with these sea trout? The truth is they are the most mysterious of all the Evitt fish (we boringly insist on calling them sea trout), defying nearly all the norms of life in a chalkstream, gripping the wild imagination of fishermen unlike any of the other species.

But why? To start with look at their life cycle. With a salmon it is pretty clear: you are born in the river, then head for the ocean and do your growing up off the Greenland coast, to return a few years later to the river of your birth to spawn and in all likelihood die thereafter. Brown trout on the other hand live more parochial lives, but with sea trout things get complicated. First, they are in every sense of fish science the same fish as brown trout. The brown trout is *Salmo trutta*. The sea trout is *Salmo trutta*. Salmon for the record are *Salmo salar*. Put at its very simplest, sea trout are run-of-the mill, bog standard brown trout who decide to run away to sea.

There are many theories as to why the sisters of Scar Boy up sticks and head for life in the salt water: lack of food in the mother river, low- or high-water years, genetic programming to ensure the preservation of the species, with some staying in the river whilst others go to sea, or simply the occasional freak event that drives behaviour. The one thing we know for sure is that more females than males head for seas because they need a rich diet to become prolific egg producers. However, before all that happens, the simple fact is that sea trout are born as brown trout, spending their first years at Gavelwood as is normal for their kind, going through the usual transition from egg to alevin and fry to parr, at which point the putative prodigal daughters will be about 5 inches long and coming to three years of age. It is here they meet the fork in the road. Take the low road and our parr moves smoothly into brown trout adulthood. Take the high road and the parr starts the smoltification process, where, like the salmon, his juvenile body undergoes physiological changes that will make the body anadromous, capable of living in salt water. Now a smolt, our fish loses his brown colour to turn bright silver, and they shoal together during the nights of March and April before heading for the sea.

You could be forgiven for thinking at this point that this is just a feral brown trout that has decided to adopt the salmon lifestyle, but sea trout are not impelled, or do not have the instinct of salmon, to cross the Atlantic and spend years at sea. In fact some sea trout spend no more than a few months away from the river, barely venturing further than the estuary before returning home. Others, on the other hand, travel hundreds of miles up and down the British coastline gorging themselves to become as much as ten times larger than the siblings they left behind. But like salmon they have that natal urge to return to the river of their birth, and come the warm nights of late June our ears become tuned in for signs of their arrival.

The best sea trout fishing takes place at night, somewhere from ten o'clock to two in the morning, and there is a saying with sea trout fishermen that you should never cast a line until the grass on the opposite bank loses its colour. So as the dusk turns to night gloom I find myself drawn to Pike Pool every Midsummer's Night in the hope that they have arrived. The fascination with sea trout lies, at least in part, in that you know, or think you know, that they are there but you can't damn well see them. You could spend all day (some have) peering into the depths of Pike Pool to catch a glimpse and nothing. Sea trout are like silent-running submarines, arriving unseen and unannounced, at the time of their choosing, using the cover of the darkness to make their run from the sea to Gavelwood. Like salmon they are making the return to spawn, but arriving as they do as early as June they are well ahead of schedule and will hang around for the autumn. So it begs the question, if they are around for so long, why don't we see more of them? After all, the river is hardly the ocean, and every other fish will make itself known from time to time, even the salmon.

The fact is that sea trout, unlike the brown trout they once were, are nocturnal, preferring to move at night, and the first time I know they are back is when I hear a 'splosh' as they leap from the water somewhere out in the darkness of Pike Pool. It's an electric moment; you rarely see the fish itself, the noise drawing your eyes through the half-dark to catch a glimpse of a few fading ripples and white waves. The larger the splosh and the larger the disturbance to the water the more the imagination races. Huge sploshes must of course mean huge fish – after all why not indulge that wild imagination? There is something strangely alluring about fishing for sea trout, as it defies most of the trout fly-fishing truths. First, you do it at night. Second, you are casting blindly to fish that may or may not be there. And third, all those precious rules about matching the hatch and delicate presentation, plus stealth and concealment, promptly go out of the window. There is something of the lottery about sea trout fishing that gives it an edge. Zipping out that fly line into the dark, the plop of the fly into the water the only indication of a cast well executed. The water catches the line, tautening it in your hand as the current sweeps the line across and down the pool. In your mind's eye you can see the fly twist and turn in the water somewhere out there in the dark, a few inches beneath the surface. What does the sea trout see in your feeble feather and fur creation? Does the phosphorescence trail the fly creates through the water from the rays of the moon trigger some memory of the ocean? A sand eel perhaps, or a baby cod? It must be a memory thing for the sea trout, who like the salmon, doesn't feed in fresh water.

Whatever the reason the fish might take my fly, there is something satisfying about the logic of fishing Pike Pool, or any sea trout pool for that matter. I start at the top, moving a little downstream with each successive cast so that I gradually cover the entire pool with the sweep of the fly. With each new cast springs new hope, regardless of what has gone before. Occasionally a splosh out there in the dark will tempt me to break the rhythm, but generally I stay with the plan. In the dark everything comes down to touch, there is none of the visual of dry fly-fishing. The fingers that grasp the line are my antennae, the first indicators of a fish. The more you cast the more you get the feel of the water, distinguishing between being momentarily caught on weed or the tentative pull of a fish tweaking at the body of the fly but not the hook.

At the end of the pool I pause. The grass on the far bank has now lost all its definition and the moon slides out from behind the clouds to illuminate Pike Pool, giving enough light to show the roots of the alder trees on the far side, stark black above the silver of the water. This is where I need to cast. Sea trout like the gloomy shade and undercut beneath the trees, scooting out under the cover of dark to frolic and chase around the pool. With a new plan I tie on a new fly and head back to the top of the pool. In the light of the moon I am no longer casting in the dark, and with each new cast I land my fly as close to the roots as I dare. As the line hits the water I lift the rod tip, raising the belly of the fly line out of the water and rolling it upstream as if doing a backhanded swing of a skipping rope. This upstream mend momentarily pauses the fly in the slack water beneath the roots, enough time I hope to catch the attention of the fish before the current snatches the belly of the line. As the flow moves the fly downstream I strip the line back towards me to take up the slack. Cast, mend, strip. Cast, mend, strip. Time after time I repeat the action, my fingers alert to the tiniest twitch, my eyes tracking the line that shows up black against the gleaming, moonlit surface. I am watching for that moment when the line stops moving, the moment when the sea trout has taken the fly in her mouth but applied no pressure. Cast, mend, strip. Cast, mend . . . suddenly the line stops moving. I strip hard and my fingers feel the pressure. Up comes the rod tip and down goes the line, surging into the deep water under the roots as the trout reacts. The fish pulls so fast that the fly line burns my fingers and I feel the bang, bang, bang as she hammers her angry head against the line and hook.

Abruptly the anger goes out of the line; not so much a slack line but an unmoving one. I know what is next. The trout has tried flight, she is now going to try fight, and comes out from under the roots into the middle of Pike Pool, where she shoots into the air before crashing back into the water. Emboldened she turns, tears downstream and in doing so rips more line through my fingers. I hold the rod tip up high and let her go, go, go, taking all the line she wants. The further she goes the more the pressure of the line in the water slows her progress, and at the tail of Pike Pool she pauses as the water shallows. Her instincts are conflicted. More flight seems the obvious choice, but shallow water equals jeopardy, and whilst she weighs up the options I take control, pulling her upstream against the current towards me. With long, smooth pulls of the line she gets closer to me as I try to keep her head on the surface to disorientate her. A few times she shakes her head in an attempt to free herself from the hook, but the fast, rhythmic movement of my pulls seems more calming than frightening.

When she is close to the bank I take the line in one hand and throw the rod down with the other, lying on my stomach over the edge to pull the last few yards of line in hand over hand until I can slide my fingers down the smooth nylon tippet to grasp the hook with my thumb and forefinger half in, half out of her mouth. For a moment we stare into each other's eyes. I have no idea what she sees; it is probably best to not even guess. But I see a bright, silver trout honed by life at sea. This is a fish that is as fit and as buff as any creature on the planet will ever be. It is a body with thousands of sea miles on the clock. She has fought tides. Evaded seals. Run the gauntlet of nets. The odds of making it this far are so slim as to be of lottery jackpot proportions. But she has done it, and with a twist of my hand the hook is out. For a short moment she doesn't move, not sure that she is free, but as the current moves her away from me she dips her head and is gone.

After all the commotion of the fish the meadows and the river suddenly seem very silent. Oppressive, no, but eerie, yes, as the whole scene is bathed in bright moonlight. It is an odd thing, but we have a few nights every summer when the moon shines as bright as the sun, casting sinister shadows and illuminating a night-time landscape that is nearly as well defined as a day-time one, just monochrome not colour. Sitting on the stile, clipping off the fly and putting away my fishing stuff I feel very much like a spy. I am in place at a time I shouldn't be, but to leave seems the wrong thing to do. As the minutes go by my ears attune and the silence gives way to the night-time life of a river. Somewhere behind me I can hear the cattle grazing. They never seem to stop. Their tongues twist around the stems of the meadow grasses and I can hear the stems tear as they pull them off the turf. Their steady rumination, the scrunch, scrunch, scrunch as they chew the cud, is the signature tune of the water meadows, the sound of the summer that is punctuated from time to time by sharp coughs and strangled exhalations. Occasionally I feel the vibration of their footfalls radiating through the damp ground. All in all the presence of the cattle provides a comforting rhythm, a sign that all is right with the world, and as long as they are content to graze unconcernedly I know nothing bad can happen to me.

The bats are still around, swooping and twisting. Like the owls, night time, be it pitch-black or white moonlit like tonight, creates no problems for them. They hunt entirely with noise, emitting high-pitched sounds, inaudible to us, that bounce back off their prey to be picked up by their supersensitive ears, which guide them in to the kill. And kill they do. Bats work at night not only because it suits their hunting modus but because they need the fields to themselves. Competition from birds and all the other creatures that feed on insects is most unwelcome, because every bat needs to daily consume one third of its body weight. That is a lot of insects, hundreds in fact, so Gavelwood is a night-time killing meadow of epic proportions as the bats patrol the layers of air from the water surface to high in the sky for hours on end. Occasionally I will see one break the surface of the river as it drinks mid-flight, dipping its mouth into the water as it makes the rapid fly-past. But eventually as the heat goes out of the evening air they become fewer and fewer, returning to their daytime roosts as the insects gradually stop flying, settling down for the night on the grass and bushes safe from sonar contact.

Way upriver, echoing on the night air, I hear a slide and a plop, followed by the same again a few seconds later, then nothing. I know exactly who this is and what they are about. It is the otters. They treat the river like a highway, travelling dozens of miles in a night if they have a mind to. These are regulars, passing through Gavelwood a few times each week, so much so that they have created 'slides' on the bank, flattened grass and slick, muddy patches where their clawed paws grip the earth, so regularly do they get in and out at the same spot. Later on in the summer they will arrive with their pups, and if they do the progress is anything but silent. The pups are incapable of getting in and out of the river quietly. They either tumble from the bank, splashing down in the water, or thrash their shorter bodies first against the water and then the bank vegetation to get purchase to haul themselves out. And all this accompanied by a regular high-pitched eek emitted to tell the parents of their exact location every half a minute or so.

But tonight the progress is too quiet to be anything other than the parents; I can only guess that the pups are still too young for hunting trips and have been left in the safety of the holt. No doubt they will be here in a few weeks' time, but for now if I strain my ears and shut out the other extraneous night sounds I can hear the pair coming as they swim downstream in my direction. I don't bother to move. I already have a prime seat. Pike Pool is a regular destination for the otters. They enter it in unison, their heads poking up and their side whiskers dripping with water, drooping long and thick, the moisture catching the rays of the moon. As they swim, paddling with their front paws or, if drifting with the current, gently undulating their broad tail for extra propulsion, they constantly rotate their heads 15 degrees to the left then 15 degrees to the right, as if scanning the water ahead. I used to think they were on the lookout for food, but perhaps it is more about sweeping the pool, trailing those supersensitive whiskers in the water to pick up the vibration of prey below. From time to time one of them will arch its back and slide beneath the surface, popping up a few seconds later. Whether it is reconnaissance, boredom or just playfulness I do not know, but they will continue to do this as they circle the pool until quite suddenly and in complete contrast to the previous dives, which were almost silent, the pair will crash-dive beneath the surface. The hunt is on.

It is said that otters can hold their breath for up to four minutes; if they can they don't do it when hunting in Pike Pool. No doubt the effort quickly depletes their air reserves, and within less than a minute one or other will surface, spitting out air and spray in a violent, strangled exhalation that reverberates across the pool. If the exhalation sounds tortured, the inhalation sounds desperate as the otter sucks in air like a sixty-a-day smoker wheezing at the top of the stairs before plunging under the water again. Otters like to hunt in pairs, it is clearly most effective, and within a few minutes one of them will finally surface with a fish sideways in the jaws and still flapping. From this distance I will never be able to tell whether it is a sea trout or brown trout, but it is definitely not an eel or grayling, the only other types of fish at Gavelwood the otters would chase with such enthusiasm. Up onto the bank the two bound and with a deft flick the fish is turned head-first towards the otter, who grips it with the front paws ready to deal the *coup de grâce* with a bite to the back of the head. It is only when the fish is dead that the otters relax, the body lying between them as they do a quick scan around them. They spot me on the first sweep, but rather than being alarmed they seem nonplussed. That's the thing about otters; they are most definitely kings of the river valley. Nobody eats them, nobody attacks them and as long as there is water in sight their line of retreat is guaranteed. But humans they don't really compute in their world. Large, yes. Dangerous, probably not. But on the whole, maybe best given a wide berth. So, dismissing me as innocuous but best avoided they pick up the fish and disappear into the night, heading along the bank with that gambolling run that makes them appear as fluid along the ground as they are in the water.

Somewhere upstream I hear them slide back into the river, and very soon the meadows settle down into the deep quiet of the early hours. The fish have stopped moving, idle until dawn at the earliest. Brown trout feed largely by sight, so the night time closes down their ability even to take nymphs, and in truth at this time of year they will wait until well past sunrise and for the warmth of the day to bring on the hatch before deigning to show themselves. The sea trout have taken to the deep recesses for another twenty-two-hour cycle of doing nothing. After all it is a long wait for these early arrivals, weeks and months until they begin cutting redds in October, so without the desire to feed, the emphasis is on keeping their body intact until the time comes. It is fair to wonder why they insist on breaking this purdah by cavorting in Pike Pool for a few hours each night. Maybe it is pent-up energy; after months or years at sea, an attempt to throw off the confinement of the river.

Two hours past midnight and Gavelwood sleeps. There is a distinct chill to the air as the lack of cloud cover allows the heat to dissipate and a cold dew covers the grass, even settling on the wide, flat backs of the cattle that continue to graze heads down and barely moving. Everyone else is tucked up. The waterfowl are hidden in the reeds. The voles and field mice are in their burrows. The fish apparently don't exist. I am sure the crayfish are out scavenging the riverbed, but I can't see them. As I make my way along the bank, leaving a dark footprint trail in the silver dew, I hear the occasional 'gloop' as a bed of weed gets sucked down by the current and surfaces again. On a fence post ahead a barn owl observes my progress with indifference. His hunting night is over; all his food has gone to ground. Maybe he is hanging around for an unwise baby rabbit to show itself, but he doesn't even bother to move as I pass within touching distance. We both know it's over for tonight. Tomorrow is another day.

12
HIGH SUMMER

DAWN. IT RISES without a chorus over the river. Typically we don't have many songbirds and more typically river creatures are late risers. As the sun burns off the mist, the fringes along the riverbanks glisten with dew-laden spiders' webs. They are everywhere. If you had to count them you would reach a hundred within a few paces. Turn your back to the river and the entire meadow is almost white with webs, from close to the ground to high up on the tall grasses. Even the gaps between the strands of the barbed-wire fences are occupied. And on every one is a spider that crabs across to consume the night-time harvest.

As a fly-fisherman, so by default a sporadic entomologist, I stoop to see the captured. There is a pattern –

the further I am from the river the more the webs hold terrestrials like black gnats and flying ants who don't need the water as part of their life cycle. Sometimes I'll come across a web that is almost shredded where a big daddy-long-legs has fought a hard but ultimately useless battle to free himself. But I don't give these webs much time; it is the ones by the river that really concern me.

Fly-fishing is all about reducing the odds. Somewhere out there are over six thousand different artificial flies, and your job, if you want to be successful, is to hone down that number to one single fly to tie on your line for any given moment. Now you could be entirely random with that choice. Or base it on what someone has reported in the fishing book. Or you could use the fly that worked the last time you fished. Or you could just look at that spider's web, identify what hatched the previous evening and conclude, probably rightly, that if it hatched the chances are a fish ate it. As I examine the webs closest to the river my interest is more than purely entomological; tonight I have every intention of fishing the evening rise, a magical time between dusk and dark when the river can boil as the fish go on a feeding frenzy the like of which we will not have seen since the mayfly.

For all the excitement, it is a time that can be maddeningly frustrating for the angler as the fish rise here, there and everywhere but consistently ignore everything you offer. In the gloaming it is impossible to accurately identify the insects on the water. A torch is no answer as the beam will immediately spook the fish, so you are reduced to scooping up a handful of water and myopically staring at the fly. At best you can hope that the silhouette will give you some clues, but the truth is fly identification is hard enough in broad daylight, so you will likely be none the wiser. But in the cold, unhurried light of dawn with hours in which to contemplate your tactics, the spider's web is the book of knowledge, and two inert figures tell me enough for the evening to come.

The earliest risers at Gavelwood are always the swans. However early I am, they will be there before me, wreathed in the mist, heads down under the water, picking away at the choicest shoots of crowfoot which are close to the riverbed. For a bird that rarely flies and spends nearly all its day gently paddling they seem to have a prodigious appetite. Maybe there is not much protein in their food of choice, and with that big bulk to fill maybe they have no choice either. The water voles will be out in force as well; this is clearly their favoured time of day, navigating up and down the river, weaving along the edge of the rushes, ready to dive for cover at the slightest sign of danger. The adults are in the prime of their life; their fur shines as if oiled and everywhere they swim it is with great vigour and purpose, be it for finding food or materials for the burrow. By this time in the year the vole population at Gavelwood has pretty well exploded, getting close to its peak for the year as the four or five litters born to each pair since March reach maturity. For such cute little mammals they are fiercely territorial during the summer, spreading out to mark their patch, though in winter when the conditions get tough they will become far more collegiate. But for now the occasional scrabbling spat will be resolved by one swimming the width of the river by way of giving up, whilst the victor mounts a handy section of reed, sits on his haunches and indulges in some nonchalant fur-cleaning.

Gradually as the sun gets higher in the sky and the morning warms, the river starts to wake up. Early on the occasional fish will lazily suck something down off the surface, maybe a spent fly from the evening before, but generally the fish come out from their night-time resting places to take up position in the open current, confident in their own mind that the food will be along all in good time. When the first hatch of the day gets the attention of the trout, it is not so much a hatch but a continuation of the day before, when a few blue-winged olive duns, the fecund females mated and ready to lay, lift off from their night-time perches. A few at a time, they drift off from the sedge grasses or drooping comfrey, or out from under the alder tree leaves, catching the thermal warm air as it transports them 10, 12 feet into the sky before they think better of it and hitch a ride down onto the river surface on a cooler downdraught. In the morning sun, with no breeze as yet, the surface has the texture and colour of blue mercury, and even the infinitesimal weight of the blue-wing creates a little indentation on the water as she settles to deposit her eggs. Every one that lands stands out to me as clear as a ship on the horizon, nodding gently with the motion of the current, and it has to be certain that if I can see them the ever-vigilant trout will not miss out. Doubly worse for the busy egg-layers, olives are some of the trout's favourite foods. And sure enough, with minimal effort but certain intent, one of the waiting trout tilts his head upwards, computes the position of the insect, adjusts the angle of his side fins and lets the hydrofoil effect of the current carry him upwards towards the surface. With practised ease he drifts downstream at the same pace as the olive until his mouth is precisely below its tiny body, at which point he breaks the surface with his nose, opens his mouth, and as the water rushes in it carries the olive into his gullet. As the trout returns to his original resting spot I can see the slight convulsion in his body as he swallows the mouthful of water and the fly it contains. From start to finish the whole thing took maybe five seconds. To me it seems much longer, and despite many thousand repetitions before my eyes it never fails to put me in awe of just how perfect nature can be sometimes.

But human life intrudes on nature, and I know what I have just seen is the absolute best of news for today's anglers, so I head to the Drowners House to share my secret. The blue-winged olive has a status in fly-fishing unmatched by any other fly. Iconic pretty well sums it up, so when I drop what I have seen into the conversation thoughts of a lazy start to the day are abandoned by all, and in a flurry of rods, flies and general fishing kit they head off to the river, whilst I take my cue to head in the other direction to see what's happening on North Stream.

Truly the best thing you can say about any restoration is that it never happened, not in the sense that it didn't actually happen, but that six months or a year later you would never know the work had been done. By that measure North Stream is almost a complete success. Looking upstream from Bailey Bridge I struggle to recall how dismal it looked the first time I saw it; barely flowing, overgrown, in perpetual shade and impenetrable for most of its length. If you were being kind you would have called it a neglected ditch. But today the scars of our work have healed and it is what we wanted it to be: a clear, fast-flowing offshoot of the main river that makes the Gavelwood meadows a better place to be for the river valley creatures, whether they live on land or in the water. The bankside plants are better for it too. Gone is the scrubland of brambles and nettles, and in its place an apparent hotchpotch of wild flowers that decorate the fringe with a whole palette of colours, growing to different heights, some upright and tall, others tumbling down into the water giving the stream a ragged edge that takes the harshness out of the otherwise straight bank.

The tallest of the plants that proliferate along the banks, about shoulder height to me, is the cow parsley, *Anthriscus sylvestris*. It grows like a small tree, topped by a canopy of large white flowerheads, each one the span of your hand, that combine to create a sort of dome. For the fishermen it is something of a nuisance, its height perfectly set to catch a careless back cast; the thin tippet at the end of the fly line always seems to wrap itself around the buds and the hook to embed itself in the woody stem. Add to that the sticky nectar that envelops the whole flowerhead and glues up my fingers so that after five minutes of patient unravelling I sometimes wonder why we tolerate the cow parsley. But somebody loves it, because in June and July the flowerheads will turn almost black, covered by tiny flies which I guess are feeding on that sticky substance I find so annoying. What these flies are, I have no idea. They don't come onto the fishing radar and look suspiciously similar to the ones that gravitate towards fresh cowpats. But I guess we shouldn't hold that against them, and somewhere in the balance of nature they have their place, so the cow parsley stays.

The remarkable thing about the plants along the bank of North Stream is the huge diversity that seems to have sprung from nowhere. Our sole effort has been to remove the scrub growth, open it up to the light and keep fast growers like nettles at bay. If I had wanted to create such a beautiful wild collection on purpose it would have been a mighty task, but nature has done it for us in the space of a few months. The purple of the loosestrife, the white of the comfrey, the yellow of the fleabane and the pale pink of the hemp agrimony all cascade this way and that. Even in the wet margins of the stream there is colour where the dainty bright blue flowers of the water forget-me-not provide cover for the nymphs and baby sticklebacks.

My self-congratulatory musings are broken by a small sound from the edge of the reeds. It is a barely audible slurp that comes and goes in a moment. A minute or so later it happens again. For years this sound confounded me. By every norm it should be made by a fish, but I could never see one, and it was only by luck that I found the source of this mysterious event. Wading upriver one day I heard that selfsame slurp, so I stopped still and stared at the reeds where I thought the sound had emanated. Nothing happened for a while, and then the surface of the water broke, slowly revealing an inch-long damselfly nymph, crawling up a reed that it used as a ladder to get from water to air in preparation to emerge from his or her aquatic body. But for at least this one nymph, being a nymph was the best he was ever going to achieve, because directly behind him, up slid a silvery black, snake-like head that slurped him down in a trice. The stealth hunter was an eel, and by just about any measure the story of where this eel has been and where it will be going trumps even that of the Atlantic salmon, spanning three decades, nearly ten thousand miles and a return journey to the Bermuda triangle.

When it comes to age, eels are the oldest residents of Gavelwood by some considerable margin, though they are almost the most rarely seen. For a furry creature like a vole, two years are a lifetime; otters might make a decade at most. Of the fish, a trout of seven is old, a salmon that makes two or three returns might just get into double figures, and often a pike may live to a ripe age beyond that. Waterfowl that get to three years of age have done well; for more wily residents like the kingfisher five years plus might be normal, and with the swans a dozen to fifteen years is a good age to aim for. So how is it that the eels I spy on a sunlit July morning, in a few inches of water of a tiny stream in an obscure corner of the English countryside, are six thousand miles from their place of birth and have outlived all the other chalkstream creatures by some considerable margin?

To start with the European eel, *Anguilla anguilla*, is not like any of the other chalkstream denizens. It is catadromous, which is to say that it lives most of its life in fresh water but must head for the sea to spawn. There are in fact only a handful of fish (yes, an eel, despite appearances and many other oddities, is a fish) that organize their lives this way. Far more common is the life cycle of that other chalkstream adventurer the Atlantic salmon, who is anadromous, doing everything in reverse to the eel by living at sea for most of the time but needing a return to fresh water for spawning. But of the two it is the eels that outnumber the salmon hundreds to one, and though largely forgotten today, they used to be one of the most sought-after fish in the river.

It is hard to know exactly where to start with the eel, so elongated and complicated is their life. Over two or three decades our eel will take in everything from muddy ditches, treks overland, ocean voyages, numerous hazards and predators, plus a pleasant sojourn in a chalkstream. But maybe this summer moment is as good as any to pick him up as he enters the last two years of his life where the only certainty is a lingering death at the end.

On the Evitt and just about every other chalkstream, eels are a common sight by late summer. After anywhere up to twenty years of living in a muddy pond or ditch they have slithered across meadows, through woods and whatever gets in their way to find the same river they swam up from the ocean all those years before. Though most definitely from the fish family, the eel has two distinct characteristics that set it apart from the other fish and allow it to migrate across land. First, and most obviously, with its asp-like sinuosity it can travel inconspicuously and without much effort over land. That said, the herons at Gavelwood soon latch on to this bounty, leaving the river for a while in high summer to patrol the far reaches of the meadows. However sly an eel might be, the patience and eyes of the heron are better. Second, the eel has the ability to retain water in its gill cavities, keeping the delicate bronchial folds afloat, allowing it to breathe as if still submerged. If you ever see a live eel on a fish counter you can see this for yourself. Look out for the slight bulbous swelling either side of the throat, which is obvious in much the same way that a hamster stores food with puffed-out cheeks. Add to that a myriad blood vessels close to the skin surface that let the eels suck oxygen from the air via the water on their skin, then, just so long as they can stay damp, being out of water holds few perils. Sadly for the eel, this remarkable ability to survive out of water was to make it the ideal candidate for the live eel trade which in its day was big, big business.

At Gavelwood we usually find out that the eels are back by accident, as we clean the fish at the end of a summer evening, lobbing the guts into the river for the scavengers like the crayfish to feast on. It usually takes a few minutes, but eels are like sharks – they can sense blood in the water. In the deep pool outside Drowners the guts tumble on the gentle current or get caught in a back eddy. Eels are hard to see until they are near the surface, but the pale pinky-white guts of the fish are not, so that's what we watch until they suddenly start to move as if grasped by a hidden hand to quickly disappear from view. If I keep chumming the river with offal, more and more eels will appear, to the point that they grab the guts from the surface the moment they hit the water. Anyone who has never seen this before will always watch with wide-mouthed amazement. It can be a real feeding frenzy; the violence and speed with which the eels take the bait is terrific. If eels were of any great size you might think twice about putting your hand in the water ever again.

But catching eels one at a time with bait is no way to make a fortune; industrial-scale eel traps were the answer, which exploited the need for the eel to migrate downstream en masse on particular nights of the year prompted by the phases of the moon. We don't have them at Gavelwood, but relics of the eel traps remain on the Evitt. Essentially a bridge was built across the river, preferably on a section that was no more than 3 or 4 feet deep, with stout posts to support it every 9 feet. Then between the posts was lowered into the water a frame that held three fyke nets, which once in the current billowed out below the bridge like windsocks. So as the eels swam downstream they were directed into the mouth of the fyke net, which narrowed towards the end where it was tied off; pressed by the current, the eels were trapped in a squirming ball of their fellows.

Today, with refrigeration and fast transport, it is hard to imagine just how sought-after fresh fish was in rural communities far from the sea, but the hundreds of tons caught in the Evitt every year had a ready and eager market. At the height of the summer when the eels were most plentiful (and most other fish and meat perished in a day), the fact that they could be transported and sold live put them at a premium. And when the steam railways connected the country to the industrialized towns the scale of the eel harvest ratcheted up further, peaking at the end of the nineteenth century.

A hundred or so years on the eel has fewer hazards to negotiate, yet even so the population has collapsed by 80 per cent in three decades. If fishing alone was to blame the solution would be easy, but with each passing year I see and hear fewer eels on summer evenings. Pollution, a particularly harmful parasite from Southern hemisphere eels, changes in the oceanic currents, and closer to home the loss of water meadows, are all contributing to the decline. Despite that there are still commercial eel nets left working in the Dorset and Hampshire estuaries, laying out fyke nets overnight eking out a living for the eels they need for smoking, but on the whole the yellow eel, as it is known at this stage in his life, heads downriver unmolested, feeding as he goes. They are, as the fish guts demonstrate, utterly omnivorous; dead fish, live fish, invertebrates, snails – you name it, they will eat it. They are driven by the imperative of this last opportunity to build up their body reserves, because in a few weeks' time they will enter the sea, never to feed again. But though they cannot feed they still have nearly two years to survive, which they will do by eating themselves from the inside out.

What fascinates me about the eel that slides past me under the Bailey Bridge on a route that will take him past all the same waymarks from years earlier, and through one of the relief hatches at Middle Mill and on to the sea, is how very little we still know about the life and habits of *Anguilla anguilla*. All sorts of theories abound and assumptions are made, but in total we perhaps know a quarter, or maybe a half at best, about how the eel lives its life. But this much we do know, that by high summer all the eels that sense their time has come are vacating the muddy ponds and ditches that they have called home for twenty years or more and, guided by the smell of the chalkstream water, thread their way through the grassland of the damp channels of the meadows and into the Evitt. The river journey to the sea of 30 miles or so is a trifle compared with what lies ahead, so they dawdle, feeding where they can whilst keeping a wary eye out for their few predators like otters or scrabbling down into the gravel for cover.

As the fresh water gives way to salt the body of our eel starts to change again while he runs the length of the estuary before turning west, hugging the southern English coastline for a while. The yellow skin starts to turn bright silver and the eyes grow huge and black. It was assumed for years that the eyes enlarged for better night-time vision, a safer time for the eels to swim close to the surface, but maybe this is not the case. As the English Channel merges into the Atlantic Ocean our eel should by all accounts slightly alter his course for a straight shot towards the Sargasso Sea that lies about 1,000 miles off the Florida coast, southwest of Bermuda. But he doesn't. He turns south through the Bay of Biscay and heads down the African coast until he picks up the North Equatorial Current, on which he will drift pretty well due west for the next year to eighteen months until he reaches the huge morass of drifting sargassum weed that is the Sargasso Sea.

But travelling on an ocean current over which he has no control creates problems for the eel, as with every day that passes he is getting closer to sexual maturity. Somehow he has to time his arrival at the Sargasso Sea to coincide with his sexual peak, because with a body that is slowly rotting from within there is no margin for error. It seems those big eyes have nothing to do with night-sight but everything to do with swimming at great depths, as the eel goes as deep as 3,000 feet in search of the best currents to help him on the journey to the Sargasso Sea. Maturing too quickly? Get down to those lower temperatures and slow down your metabolism. Arrested development? Move up to the warmer water closer to the surface for a while. One way or another, the eel arrives ready to mate.

I would love to tell you that the eels have a seagoing version of the redds and a ritual as poetic as the trout and salmon, but sadly no. The truth is nobody is exactly sure how eels breed. The Sargasso Sea, at 700 miles wide and 2,000 miles long, is still one of the great unexplored areas on the planet. To the best that anyone can tell the female lays her eggs in the water, into which the male releases his sperm, the tiny fertilized ova attaching themselves to the weed during incubation whilst the parents simply die. With anywhere between 2 and 10 million eggs being laid by each female, it has to be a fairly random process, but how long the incubation lasts we can only guess at. What is known is that they hatch as *leptocephali*, leaflike larvae, and countless millions of these tiny, transparent, immature eels hitch a ride back towards Europe on the Gulf Stream, taking the direct northerly route that their parents avoided.

It is a fairly aimless journey that can take anything from a few months to a year but once the eels, now about 2–3 inches long and known as glass eels, scent the smell of a river they are galvanized into action. Prodigious numbers converge on the river estuaries, and nature's tactic of providing millions so that a few may survive is amply demonstrated, whilst the birds and fish gorge themselves on this annual feast that comes their way in May and June. Man, of course, learnt not to let this windfall pass by without harvest. In Victorian times the coastal fishermen of Devon and Cornwall netted such vast numbers that they were sold by the cartload, to be fried into so-called 'elver-cakes'. Heaven knows what they tasted like, but the description at the time as 'having a peculiar appearance from the number of little black eyes that bespangle them' does not make me want to grab for the frying pan.

But assuming that our elver, now starting to get darker in colour and lose his translucence, makes it past these dangers, he finds himself compelled by millions of years of evolution to migrate inshore using the Evitt as his highway, hauling himself out to wriggle across the fields to make his home in a muddy ditch or stagnant pond. An eel can pretty well eke out survival in any habitation, growing according to the season and availability of food. Even extremes of weather don't faze them. In dry spells he will burrow deep, and in freezes will get frozen with no apparent ill effects. This was famously discovered by the nineteenth-century French zoologist

Eugene Desmarest, who kept a pet eel for thirty-seven years. However, during one particularly harsh Parisian winter Desmarest arrived at his laboratory to find the eel frozen in the pottery pan in which it lived, so brought it back to life by thawing it with tepid water. In fact there are plenty of folk tales of frozen eels mistakenly gathered in firewood and 'miraculously' springing to life when piled up by the hearth.

But for all its ability to survive, our eel is no fast grower. The usual span in the muddy lair is somewhere around twenty years, which by most measures is a long time, though up to eighty-five years has been recorded, putting the apparently mundane eel right up there in the animal kingdom longevity stakes. Fortunately the years of life are no guide to size, as the eel will rarely grow to weigh much more than three pounds or be longer than 3 feet, otherwise he would be of anaconda proportions, devouring everything in his wake. That equates to an annual growth rate of just over an inch a year and four ounces of weight. In truth an annualized average is probably deceptive, as our eel grows in spurts, putting on plenty of weight in the good years and just surviving in the bad.

In the early evening as the sun dropped in the sky and the smoke from the barbecue drew me across the meadows towards the Drowners House I pondered on where to rank eels in the chalkstream hierarchy. On the one hand they are omnipresent, somewhere out there in their thousands, biding their time for that moment, years or even decades ahead, when they will start migrating. Until that moment comes we hardly see them, let alone have any interactions with them, so they hardly feature on the chalkstream radar. Two centuries ago, when they represented big money, I know I would have felt very differently, marking out these summer nights for a profitable eel harvest. Back then the brown trout that are everything to us today were held in very little regard, so I guess every creature will, in the end, have its day.

Nobody at Drowners is in a big rush to go fishing; tonight is as near perfect as it will get. That sort of evening when the air is still and humid. The type of night when you throw off your sheets and blankets, open every bedroom window and lie on top of the bed wishing that you had air conditioning. Nobody rushes because fishing the evening rise is as much about anticipating it as doing it. It could happen any time from when the sun dips below the horizon to when a cold mist rolls over the meadows. Or it could not happen at all. From Drowners we have a pretty good view up and down the river, so everyone is content to loll about with bottles of beer and desultory conversation waiting for the first sign, be it a splash of a messy rise, a tiny dimple, or the 'gloop' as a sizeable fish sucks down a spent fly. Our ears grow more attuned as our eyes grow less useful.

It is part of the mystique of the evening rise, that brief moment of time after dusk but before darkness when the fish go on a feeding frenzy, gorging themselves on the insects as they die or lay eggs in the surface of the river, that it does not happen every evening. The allure is not so much the catching, but the manner of the catching. In truth, for all the effort and dud evenings it is one of the least efficient manners of catching a trout. But we persist because one good evening, even one good fish, makes all the failures worthwhile.

Tonight the omens look good – bright sunshine all day and now as dusk advances, a hint of humidity. We are banking on the fact that the trout are hungry after a day in which they hardly fed, not by reason of the heat but simply because trout don't have eyelids, so looking up hurts. However, the feeding habit is well ingrained at this time of year, so really we are waiting them out, knowing or perhaps more realistically hoping, that temptation will get the better of them once the sun has set and the insects, dead or alive, get stuck in the water film in the close, humid air of dusk. So wait we do, as the heat goes out of the day and the light fades.

'Upstream or downstream?' I ask, uncurling from a bench and picking up my rod at the sound of the first plop. 'Upstream I think,' replies one of the others joining me as we creep along the bank ready to make the first cast of the evening. More or less in position, we wait for the fish to show again, and sure enough 10 yards upstream and across, below the opposite bank, the silver rings of ripples show where the trout has taken a fly in the surface film. 'What fly?' asks my companion, as if anyone can really tell in this sort of light. 'Sherry

Spinner,' I say with absolute certainty. He looks at me with mock incredulity so I say, handing him the rod, 'Here, you have first cast.' My choice of fly has nothing to do with what we have just seen but everything to do with the cobwebs from the morning that were thick with blue-winged olive duns. Logic tells me that the duns, even though ready to mate, will have hated the hot day as much as the trout (and us for that matter), so will have waited it out for the cool of the evening to make that final transformation into mating spinners and egg-layers.

Kneeling behind the bank fringe for cover, my companion makes short little flicks of my rod to pay out the line, dropping it to the surface to check the fly is floating OK. Satisfied, he picks up the line, gauges the distance to the fish, and after two deft trial casts in the air then lands the line on the water with the third, the fly alighting gently on the surface about 2 yards ahead of where we estimate the fish is holding. In the gathering dark it is surprisingly easy to see the line and the fly, which show silver against the inky dark of the river surface. It is made easier still when suddenly the water erupts around the fly and with a whoop of joy he swiftly raises the rod tip to hook the trout. But this trout obviously knows a thing or two about being hooked, for it heads to a weed bed that in the dark the angler cannot see and promptly throws the hook. One nil to the fish.

Leaving him to curse, but silently congratulating myself on fly choice, I move on upstream as the river starts to come alive with fish rising here, there and everywhere. This is a true evening rise and I know that it won't last long. The temptation is to cast like mad and fish like a man possessed, haring up and down the river to cover every fish. However, the usual rule of fly-fishing applies: pick your fish, cast once and make your one cast your best cast. I observe this rule at least in part and soon have three fish to my name, one fly lost in a bush, and both hands red and itching, having managed to sting them in a clump of nettles when releasing one of the flies.

Holding up the fly to the sky and squinting to check it, I realize that in my concentration on the fishing we have gone from dusk to dark and the fish have stopped feeding. As the minutes pass I will a fish to rise, but with no new rises and the mist rolling in over the river I know in my heart of hearts it is over for tonight. For a while I hang on, hoping for just one more chance at one more fish, but when the heat of the day ebbs out of my body and a little shiver goes through me I take the hint. The rise is over, it is time to go, but the evening had lived up to its promise.

13
THE ENGLISH SAVANNA

MAKE HAY WHILST the sun shines goes the saying, so in early August we do exactly that. In pure farming terms we are later than most, in fact as much as six weeks later, but cropping the water meadows of the mixed grasses is more than just creating forage. In sheer practical terms, the meadows at Gavelwood were never designed for modern-day farm machinery that likes flat, geometric fields with easy access. Not only do our field boundaries follow the erratic course of the Evitt, but they are cut into odd shapes by the course of North Stream and Katherine's Brook. Flat? Well, from a distance they might look flat, but head across them in a straight line and you will find that they are anything but, as the ground rises and falls in regular undulations that give the walker the slight sensation of surfing a shallow wave pattern, with maybe a 2-foot fall from peak to trough. These are the now redundant ridge-and-furrow water-meadow channels that radiate like ribs from the spine that is North Stream, and in winters and springs past would have flooded the fields.

Today, even in the comparative dry of August, the troughs are still wet and the tractor driver has to avoid the wettest sections or else get stuck, which is no bad thing, as a few uncut sections provide a useful refuge for plenty of the creatures. Occasionally, through misjudgement or just sheer bad luck, the tractor wheels will break through the surface and gouge huge ruts in the soft turf. Exposed, the sward that covers the meadows, which looks so permanent when viewed overall, is surprisingly fragile, the turf and roots no more than 2 inches deep. The soil beneath is deep black, more like wet clay, and apparently devoid of life other than a few earthworms; it squelches when I kick the sods back into the ruts and stamp them down to repair the damage. Left for a few minutes the ruts would soon fill with water, which is a gentle reminder that this land was, is and always will be a floodplain where the water table is constantly present, just a few inches beneath our feet. Once in a while I will spot an eel squirling in the water of a rut; whether we have disturbed his long-term home or he has broken his journey across the meadows back to the river for a revitalizing dip I cannot tell, but either way his vigour suggests he enjoys the feel of water on his skin.

Regardless of the practicalities I try to leave the hay-making as late as possible to preserve the homes and hunting grounds for the valley creatures. The field mice thrive in the thick base of the grasses. The bolder water voles range away from the river in search of seeds. Curlews, lapwings and the very occasional skylark make their nests and raise their young in the meadows, the damp ground providing all the food and cover they are ever likely to need. The river insects spread out all over. For the hunters like the bats, owls and herons it is a veritable English savanna. But in the end the cut must be made; removing what is now largely dead will bring on a new sward that will grow fast with the warm summer air and damp soil to herald a new phase in the annual cycle of Gavelwood.

With the long, hot dry days there is definitely a sea change in the pace of life for everyone. Gone is the frenetic activity that goes with courtship, hatching eggs, caring for the young and competing for territory. August is a time to consolidate. With the exception of the invertebrates that carry on regardless, just about everyone to be born this year has been born. That tricky phase from egg or infant to tiny creature has been successfully negotiated. Certainly there are plenty of hazards ahead, and your chances of being alive twelve months hence might still be slim, but to come this far is a victory in itself. August in the river valley is as close as it gets to heaven on earth for all the creatures; with warmth, food in abundance and healthy, growing bodies, there is very little bad.

The water meadows will spring back to life within days of the hay cut. Haymaking inevitably shakes out millions of seeds per acre that percolate into the turf. By day the birds endlessly peck away and by night the foragers move in. Now exposed to the sun, the wild flowers soon bloom, carpeting the turf with all manner of colours and variety; the yellow of the birdsfoot trefoil stands out, as does the meadow buttercup. The plain, stubby brown flowerheads of the ribwort plantain are everywhere, their ordinariness accentuated by the beautiful violet-purple flowers of the low-growing tufted vetch. Lie for a moment in a summer meadow to take in the huge kaleidoscope of colours. The twisting, twirling shapes of the leaves. The stalks that grow dead straight or entwine with competing plants. Butterflies will dance above you, and on every surface there will be insects, ants or spiders. And all this takes place amongst a cacophony of noise. At first you will not be able to place it, a white noise that is at odds with what should be a peaceful landscape. And then you will notice a bumblebee. Then another and then another. Raise your head to look across the tops of the flowers towards the horizon, and there will be thousands of bumblebees flitting from stamen to stamen as they collect pollen, buzzing as they single-mindedly go about their task. Don't worry about a bumblebee stinging you, they rarely do; wasps on the other hand are an altogether different story.

It is easy to think of river keeping as the perfect job, and in many ways it is. A chalkstream valley is no bad workplace and the river creatures are amiable work companions, blithely unaware of office politics. Certainly the weather is not always kind, and sometimes the work is very hard, but being effectively your own boss with a schedule of your own choosing is enviable. But without overstating the case, the work has its dangers: farm machinery and chainsaws come with risks. Working alone and unsupervised is not always a good idea. Most alarmingly, I know plenty of river keepers who can't swim! But ask a river keeper what he fears most and the answer will be one word – wasps.

Wasps have a particular liking for the meadows; abandoned water-vole burrows make the perfect place for their underground nests, and the very thing the colony needs for food, the insects that the workers collect to feed the larvae, are everywhere in huge abundance as they flourish in concert with the growth of the meadow plants through spring and summer. That would all be fine in itself if the nests were easy to spot, but the trouble is that the entrances are usually hidden, a tiny hole in the turf not much bigger than a fifty-pence piece that is often shrouded by grasses. If you are walking or fishing you have half a chance; by the time we get to late summer the activity around the nests is at its peak, wasps entering and leaving the hole by the dozens every minute from sunup to sundown, so the noise or sight of them will give you a clue as you move along the bank. In the early months of the season the nests are less of a problem, and you can pass by the entrance without a second glance. However, as the colony expands the wasps grow notably more aggressive as they commute back and forth in an ever more frantic search for food. By August it is wise to give any nest a wide berth. Unfortunately river keepers, when driving a mowing machine or using a weed strimmer, don't always get that prior warning. With the dust, noise and spray of cuttings, plus sometimes a protective face shield and earmuffs, the cloud of angry wasps can be up and around them before they even realize they have hit the nest. And wasps are vicious. Unlike bumblebees that sting as an absolute last resort and die after making that single sting, wasps can sting repeatedly, and they do. It sometimes looks comical to see a river keeper in the distance abandon his machine and run like the blazes swatting away around his head, but it is no fun really. Most keepers will tell you of the times when they have had multiple stings, and some carry injection pens in case of a severe reaction.

In general it is fair to say that the annual cycle of the river-keeper year does not contain many dangers like the wasps, but there is definitely a cadence as the months roll by that pushes us along and makes the job so special. Winter is undoubtedly the hardest. It is not just that the work is physically demanding, but sometimes the dark mornings, short days and driving cold rain can seep the will from the toughest bodies. Gavelwood doesn't give you much back, as the creatures are either absent, hibernating or hidden away. But most days there will be something that reaffirms your hope and offers a reminder of better times ahead. A bit of watery sunshine that prompts a hatch of tiny olives. A water vole that pokes his head out of the burrow, then scurries off for food. A big old kelt salmon, exhausted from spawning, that swims past making his way back towards the sea. A perched owl that looks as wet and miserable as you do, with whom you share a shrugged glance that seems to say, 'What can you do? It's winter after all.' But it does end, and all of a sudden the days are longer. The brown valley begins to get a tinge of green, and the first time the mowing machine comes out a symbolic corner is turned. All of a sudden a fishing season that seemed so distant so recently is only a few weeks away and there is a rush of preparations for the first anglers. The opening day and shortly after is a pleasant pause. It is a restoration of faith, a time when everything you work for all year has a point as the river gives up pleasure for the anglers who revel in the beauty and the sport. As May turns to June the workload multiplies; there will not be enough hours in the day or days in the week. The grass that needed mowing once in March now needs to be cut at least once a week. The miles of riverbank fringe that have been gradually getting taller without you noticing now need trimming. And every day there is something new. A stocking to supervise. A group of fishermen to cosset. A bridge to repair. The June weed cut is almost a blessed relief, as all other jobs get set to one side for this single priority. But I pay for the interlude, because as soon as the cut is over it is a question of picking up where you left off and making up for lost time. July is very much the same as June. Another weed cut, more mowing with everything still growing, and it is not really until August that there is a moment to pause for breath.

August gets a mixed press with anglers as a fishing month; some revel in the peace of the summer meadows and the uncrowded riverbanks, as many of their fellows fall prey to a family holiday that will rarely include fishing. Others throw up their hands in shock at the slightest mention of fishing in a month in which they consider the water levels to be too low or the weather too hot. I am firmly in the go-fishing camp, but the stay-aways have a fair point, and at Gavelwood August is the month when I have to box clever to keep the river in prime condition. It is the month, and September as well for that matter, when the Evitt will be visibly lower and the side streams, Katherine's Brook in particular, narrow as the vegetation grows in and the velocity of the water pushing through reduces. Common sense tells you this is going to happen; after all the summer months are the driest, so why should it not happen? There are all sorts of demands on the river. The trees along the banks suck up vast quantities. Farmers require irrigation. Homes need water. Evaporation is a significant factor, but most of all the chalk strata, the huge sponge that feeds the aquifers that in turn spring from the ground to create the rivers, are gradually deprived of water with less rainfall. That one drop in, one drop out principle of the saturated sponge no longer applies. The longer the summer goes on the more we are reliant on the reservoir of water held in the chalk, but it is diminishing with each day.

There is no need to panic; after all this is the natural order of things, and good winter rains will reset the balance, but for now I need to conserve what we have. Katherine's Brook is the first to suffer; if I follow the stream up from Gavelwood towards its source there will come a point where it is completely dried up by now. Though it is never good to see, small brooks are often winterbournes near the source, streams that flow entirely from the springs, and as the water table drops they gradually dry up. In wet years they keep flowing longer into the summer and the distance that dries up is less; in dry years the reverse. Down where we are, 6 miles from the source, the Brook has never dried up in living memory, but the effects of a dry as opposed to a wet year will have an impact. If you stand and look upstream it will be noticeable because the water has narrowed to maybe a third, or in really arid years a quarter, of the wintertime width. The channel will be delineated by watercress, snaking down the middle where it has grown out from the banks to occupy the space that was previously water.

This is nature's way of preserving life in a brook. In the same way that the crowfoot is a haven for nymphs, so is the watercress. And not just for nymphs. The snails, the shrimps and tiny fish successfully make this their home for a few months, out of the gaze of many of their predators. Even the occasional larger fish will occupy the margins where the watercress floats on the surface creating a refuge beneath. By narrowing the size of the river the water continues to flow pure and fast, still a perfect chalkstream, just reduced in size. And as a regulator the watercress is the river keeper's dream, adapting to the conditions, growing and narrowing as the summer progresses, even acting as a filter for silt and anything else bad that might try to flow into a river when it is at its lowest and most vulnerable. So my job is to do nothing. To cut back the watercress would be a disaster, destroying a habitat that nature has worked so hard to create, and so it all grows at its own pace until the autumn turns to winter and the heavy frosts kill the watercress (being 95 per cent water, freezing does it no good) and it dies back, opening the river up as the water levels return to normal.

Fortunately there is no commercial value in the watercress, so nobody bothers with it on that score. It is not the stuff, *Rorippa nasturtium-aquaticum*, that is familiar on supermarket shelves, but rather the aptly named fool's watercress *Apium nodiflorum*. I will admit I find them hard to tell apart at a glance, but lift up a clump of the fool's variety and you will see leaves that are shaped like arrowheads and grow in regular pairs along the stalk. The eating kind has rounded leaves that are totally irregular on the stalk, but if you are in any doubt apply the sniff test. Crush a bunch in your hand. If it smells of parsnip it is fool's watercress. Whether that is edible in any quantity I do not know, but it certainly does not have that peppery bite of the real thing, tasting more of old celery. There are a couple of spots around Gavelwood where the *Rorippa* grows, and a few handfuls in a lunchtime sandwich makes as good a meal as you are ever going to get. Everyone worries about liver fluke, a parasite worm that sheep and cattle carry, but we pick our cress from the fast water, well away from the margins, and haven't succumbed as yet.

There is very little I can do to make the Brook a live fishing prospect once we get into August; I will keep the pathway mown and the worst excesses of the fringe trimmed for the occasional angler who will wander up to have a try at one or two of the wild fish that have found a niche. Conversely, the fate of the North Stream lies entirely in my hands, by virtue of the amount of water I choose or am able to let in through the Portland Hatches from the main River Evitt. For most months of the year it is not something I have to think too hard about; in winter all the boards that regulate the water flowing into North Stream are removed and in the spring they will be replaced to slow it down. But by August I have an entirely different problem. North Stream was dug, admittedly many centuries ago, for meadow flooding.

The water engineers of the time cared nothing for summer flows; their interest was solely for the winter and spring. So all the gradients and structures that work so well most of the time do nothing to help North Stream as it is gradually starved of water, the pace of the flow falling away as the summer progresses. I could of course let it go wild, as I do the Brook, but that would fly in the face of everything the restoration was meant to achieve, so somehow I have to work out a solution. In an ideal world I might build an enormous current deflector to redirect the water out of the Evitt and into North Stream, but actually it wouldn't be so ideal, a case of robbing Peter to pay Paul. No, the solution lies in harnessing the power of what the river has in abundance. Weed.

The chalkstreams work so well because they are all about gradual gradients. Nothing dramatic; the fall of 2 inches over a mile is enough for that steady pace of water that creates the burbling streams that trout, invertebrates and all the other creatures love. Any more and the water would rush to the sea, emptying the rivers faster than the aquifers could fill them. Any slower and the cool, oxygenated water would soon start to stagnate. Somehow I knew back in July I'd have to replicate this fall between where the water comes into North Stream from the Evitt and then returns to the river some three-quarters of a mile downstream. So during the weed cut that month we left a bar of weed 50 yards long, stretching across the width of the river immediately below the entrance to North Stream, uncut. After a while it began to look raggedy and unkempt, looking more untidy by the day as the white crowfoot flowers carpeted the surface to the point where no water could be seen except for a few rivulets that forced a threaded passage between the weed. As August heated up, the surface weed flourished, and beneath the water the bulk of it thickened, so much so that a family of moorhens more or less took up permanent residence, building a sort of nest in the middle of the weed bed. The grey wagtails moved in as well to call it home, with their bright yellow bellies and frenetic tails that beat up and down as they hopped around searching for insects as if on solid ground.

Gradually (things rarely happen fast on a chalk-stream) the barrier the weed bar created started to impede the flow, raising the water upstream of it a fraction of an inch every day, which had nowhere to go but down North Stream. At first it was a question of stemming the wound, but as the days went by I could see the life come back into North Stream, and when the time came around for the August weed cut we were able to trim the weed immediately below where the carrier rejoins the Evitt to give the water free passage out of the North Stream, allowing it to flow faster still. Crisis over; the weed bar would stay in place until the autumn rains came, when we would cut it away.

Trout get lazy in summer. Too much time. Plenty of food. As the song goes, the living is easy, but in this case the fish aren't jumping or more pertinently rising, which is when the box of nymphs comes out from the fishing waistcoat. For some the very thought of fishing with a nymph is the folly of a philistine, the piscatorial equivalent of cheating at patience, but fly-fishing comes in all shapes and sizes. Essentially the style of fly-fishing on the chalkstreams falls into three groups: wet fly, dry fly and nymph. Each method has its detractors and supporters, but essentially the aim of all three methods is to cast, or at least put, something in front of the trout (or salmon) that looks like food. The rest as they say should be history.

The manner of the wet fly we fish today, which is mostly used for sea trout and salmon, is a long way from the wet-fly method that our ancestors fished until mid-Victorian times. Fishing scenes in paintings from the seventeenth or eighteenth century generally depict gentlemen in breeches and waistcoats holding long rods, maybe 12 or 15 feet as opposed to those nearly half that length today, facing and fishing downstream. The rods they used were not much good for casting in the way we do today, fashioned as they were out of stiff, heavy greenheart wood. It was more flicking than casting, the line made of tapered horsehair tied to the tip of the rod – these men fished without reels. The wet fly attached to the end was fished in the manner the name suggests, sunk beneath the surface with the current giving it movement as it swung across and downstream. The wet flies were not tied to imitate a particular nymph or insect, but rather as an attractor that goaded the fish into action – if it looked like anything it would be a small fish. As a way of catching fish it can't have been easy. On a windy day, getting the line out would have been a struggle and all that chalkstream crowfoot offered endless prospects for getting caught as the line swung round on the current. Add to that an outfit that was hardly what you would call sturdy: a stiff pole, no reel, and a horsehair line that could easily be snapped by a two-pound trout, which made the prospect of landing a fish, let alone hooking one, daunting. And heaven forbid that a salmon took the fly – you would be smashed up in a trice. This was the world that was as familiar to Izaak Walton of *Compleat Angler* fame in 1653 as it was to anyone fishing at the time that Queen Victoria ascended the throne in 1837, but during her reign everything was to change.

The change happened for lots of reasons: intellectual curiosity, an adapting society, advances in technology, but most of all because what came after (dry fly) was better than what went before (wet fly) for the very reasons people go fishing – to catch fish. The technology of the time, such as it was, brought innovations to the fishing community. Proper reels, rods made of flexible split-cane bamboo, braided silk lines and steel hooks became commonplace. The grand estates moved from being purely agrarian to the playgrounds of the landed gentry, the entertainment including fishing. Educated people had the time, means and desire to seek new ways of catching fish, and one of them was a successful manufacturer, Frederic M. Halford. It is often said that Halford invented dry fly-fishing, but that is as wrong as saying that Doctor Johnson invented English. What he did do is draw together the disparate practices of his day, codify a collection of fly patterns in common use at the time, and give some semblance of order to define what we now understand as dry fly-fishing. A chalkstream man through and through, Halford would recognize Gavelwood as every bit the type of river on which he based the two books, published in the 1880s, that sparked the dry-fly revolution. The creed, which demanded that the fly-fisher only fish to a rising fish with a fly that accurately imitated the insect it was rising to, swept the trout fly-fishing world. However his pre-eminence was challenged thirty years later by G. E. M. Skues, who truly did invent a new style of fly-fishing, that of using a nymph, to cause a rift in the fly-fishing community that exists to this day.

The problem for Skues was that by the time he started to write about nymph fishing in the years just before the outbreak of the First World War, the Halford way had become the *only* way to fish; wet fly-fishing was consigned to history, and anything that involved a sunken fly on a chalkstream derided. It is hard to imagine it now, but Halford became something of a superstar. Anyone who was anyone with an interest in fly-fishing at one point or other made their way to pay homage to him at his Oakley fishing hut on the banks of the River Test at Mottisfont Abbey, including by all accounts Skues, who published his first book on nymph fishing, *Minor Tactics of the Chalk Stream*, in 1910, four years before Halford's death. I have no idea what sort of man Skues was, but he must have had broad shoulders, for the opprobrium and downright hatred that greeted his style of fishing were beyond all reason. Cheating. Unsporting. Unethical. These were all words commonly used, and many river owners banned the use of the nymph, a ban that continues to this day on some rivers for all or part of the season. It was an extraordinary outcome really; all Skues was doing was applying the same logic as Halford – identify a feeding fish, but this time under the water, and then fish upstream of it using an imitation of the sub-surface insect, namely the nymph.

If anyone thought the schism would close with the death of Halford they would be in for a shock, as the disciples of the dry fly became more vociferous and determined to promulgate what they saw as the *only* proper manner in which to fish a chalkstream. Skues ratcheted up the pressure with a follow-up book, *The Way of a Trout with the Fly*, in 1921 that established the nymph-fishing technique in the same way that Halford had with *Dry Fly Fishing in Theory and Practice* in 1889. It all culminated in 1938 with an Oxford Union-style debate called by the committee of the venerable Flyfishers' Club entitled 'The ethics of nymph fishing in chalk streams'. The parameters of the debate and the direction of travel were clear in the manner in which the title of the debate was framed, but Skues, now eighty years old, entered the lions' den to put his case to the assembled members. To the surprise of none he was roundly defeated, the vote vastly in favour of dry fly as the only acceptable method. Fortunately for the August anglers at Gavelwood the result of the Flyfishers' debate doesn't hold much sway today, and techniques Skues pioneered are meat and drink for the thoughtful as the nature of a summer river changes the mindset of the fish. I said the fish are lazy – they are not really. It is rather that they adapt to what is happening in the water around them; after all they are better placed than anyone else to observe the habits of the insects they like to feed on, and those habits change as the days get hotter.

Essentially, if you are a nymph ready to hatch on a blazing hot day you will find you face something of a conundrum. You are all pumped up and ready to go, but when you reach the thin final layer of the river surface through which you have to push, it is an unfriendly place to be: unnaturally hot, sucked clean of oxygen and stretched tight. Add to that the fact that none of your fellows are hatching and you might wonder if all the risk and effort will turn out to be ultimately pointless. So you decide to wait, maybe for the cool of evening or the damp of the following morning, but one way or another with your feeding days over, so that you have nothing to do but flit around near the bottom to while away the hours. You are naturally not alone in this thought process, and as the day heats up more nymphs arrive at the same conclusion, with fewer and fewer hatching.

For the angler on the bank the river is glassy flat, barely a fly or a rising fish to be seen. But this is August. The streams of Gavelwood are truly pellucid, the bright sunshine illuminating the riverbed, the rays bouncing back off the gravel. Down close to the bottom life goes on in the perfect cool of the chalkstream water. It may be arid at the surface and blazing hot in the air, but for the trout and the nymphs everything is just as it should be. As I peer into the translucently clear water I don't need polarized glasses to see the fish, who gently hang a few inches above the stones, swaying with the current. If I am careful I can watch them unobserved, but the bright sunshine is a double-edged sword – if you can see them, they can see you. Confident though they are out in the river, eager to feed, they are ever alert to predators. The slightest shadow, even a bird flitting across the face of the sun, will send them dashing for cover to hide under a clump of weed or bank undercut until the coast is clear.

As morning turns to afternoon there develops something of a nymph logjam under the water, as the hatching grinds to a halt and the numbers multiply. The trout can hardly believe their luck. There are nymphs everywhere, and not even hidden away; freed of the necessity to feed, our little invertebrates abandon the cover of the weed beds to enjoy the open water. It is hardly a wise career choice, as the trout take full advantage, but it is fascinating to see how they go about it. Six months ago Scar Boy and his cohorts, half the size that they are today, would have been in total ferment, careering back and forth chasing down every nymph, expending nearly as much energy in the chase as they gained in the consumption. But now they are not lazy but wise. They know that if they wait the nymphs will come to them, and a simple movement of the head, or a slide of the body by a few inches left or right, and the food will all but swim into their open mouths. It is the ultimate reward-for-effort equation and the trout just revel in it. And so do the anglers, at least those of the Skues persuasion, who will snip off the dry fly and tie on a nymph.

The Gavelwood anglers treat this time of the year as one all of its own; a challenge thrown down by nature and the trout to be taken on in no uncertain terms. Gone are the waxed jackets, bulky waistcoats and rubber boots; the trout are cool and comfortable, so there is no reason why we should not be as well as the summer stillness envelops the river valley. The cattle barely move in the heat. The birds rest in the shade of the trees. The voles relax in their damp burrows. The otters sleep in their holts, as do the bats in the dark recesses of the Drowners House. Alone of the creatures beside the bank it is the damselfly that revels in the conditions as the mercury rises. Quite unlike any of the other invertebrates on the chalkstreams, with the exception of the dragonfly, which are far less common, they are by a long way the biggest insects you will see all year. If you are in any doubt as to whether it is a dragon or a damsel watch them fly; dragonflies can hover whilst damselflies cannot. But regardless of that, it is amazing how the neon blue of the damselfly body will stand out against the pink, yellow, purple and green of the riverbank plants. There will be hundreds, maybe thousands, within a relatively short stretch of river as they fly from stem to stem in search of a suitable mate. The mating itself is extraordinary, the two furling their long abdomens into a heart-shaped mating wheel. The male lands on a stem directly in front of his mate, stretching his abdomen back to grasp her by the scruff of her neck. Supported by the male, she is in turn able to curl her abdomen under her body to grasp his reproductive organs from below, and in this rather beautiful tableau the mating takes place.

What happens next is pretty well unique amongst the invertebrate kingdom (at least on the chalkstreams) as the two fly off together for the egg-laying. This is completely at odds with the mayflies, the midges, the sedges and just about every other fly you care to name, for whom once mating is over the usefulness of the male is at an end, left to die in the meadows as the female returns to the river. It might be one step up the scale from the post-coital female praying mantis devouring the male, but is by any measure a long way short of a romantic finale, but the damselfly pair redeem them all as they head for the margins along the river together. Alighting on a stem or reed emerging from the river, the female crawls down to lay her eggs beneath the surface, whilst all that time the male, at her most vulnerable moment, keeps guard from above until the eggs are laid and they can die together. Some days, as I creep upriver in the summer heat, a fish will throw himself out at the guardian damselfly, crashing down amongst the reeds. Rarely does the exuberance seem to have the reward the effort deserves, as the male flutters away unharmed, though quite how the commotion affects the female mid-laying one can only guess. Her greatest danger is really from the eels, and I can't imagine how the male hopes to protect her from *Anguilla* other than by being a lookout, so however chivalrous his intent, it is a mystery to me why he does what he does.

These days of late summer always feel to me as if nature has rolled out a giant aquarium through the valley. The silver ribbon of the river is perfectly framed by the vivid green of the banks. The water is so clear that you hardly know it is there, but its wetness varnishes each stone on the riverbed. The sun, so high in the sky, casts no shadow, the penetrating rays magnifying every thing and every movement. Even I, who think I know the river better than anyone, notice things I have never seen before – the brick foundations of an old bridge pier that must have been demolished decades ago. A line of rusted iron posts, the jagged tops just poking out of the riverbed, evidence of some long-forgotten bank repair. The contours of the riverbed showing up as if in 3D, a tapestry of green weed, loose gravel and soft silt. Standing there on the bank, with a bird's-eye view, all the life of a river is laid out before me. It is as if, for the first time in the year, nothing is hidden from sight. Every fish, every eel, every crayfish, every snail, every shrimp that moves I see. As my eyes grow accustomed to the glare even those at rest or keeping out of the glare come into vision. But best of all, every generation of trout is represented within each few square yards of water, proof that the future is assured. The tiny ones, no more than 4 inches long, the product of the winter spawning. The yearlings, now eighteen months old, are easily distinguished by the black tips to their fins and tail that contrast with their pale bodies. The third-year fish are the perfect ones, with elongated bodies, white-golden skin and bright red spots. This winter will be the first for them to pass on their genes, and with each successive year they will grow darker and plumper, but for now they are the fish that catch the eye.

But for all that I can see, the one thing I can't pick out is the nymphs. I know they must be there. The trout tell me it must be so, as they all, regardless of age, hover on the current, making the occasional small movement left, right or forward to swallow down the food which is unseen to me. This is too good a moment to pass up, so I open my fly box and stare at the array of nymphs, hoping inspiration might come for which to pick. After all, unlike dry fly, where it is all writ large, there is nothing I can see that tells me what the trout are eating, other than it is a nymph, but of what sort I do not know. In my head I run through a list of what it might be. The constant presence of the damselflies offers a clue;

it could well be a damsel nymph. On the other hand an olive nymph of the blue-wings caught in the spider webs seems logical. Common sense tells me that at this time of year the caddis nymphs will be stacking up ready to hatch. Whichever I choose, it is really through inspired guesswork and by trial and error that I will eliminate each until I find what works. Maybe it was this, the imprecise science of nymph selection, that so infuriated the Halford disciples.

Some people say you need to achieve a Zen-like state to be a nymph fisherman of the highest order, with your actions and mind totally focused on the fish and the act of fishing. That might be something of an exaggeration, but if I am in a hurry to be elsewhere or my mind is on other things it rarely goes well, and the belief that you need to see the world without distortion definitely rings true. Kneeling down behind a clump of fluffy meadowsweet for cover, I get in position to survey the fish 10 to 15 yards ahead. In terms of which fish to target I am spoilt for choice. Part of me says go for the biggest. Another part says just cast and hope. Neither option seems very Zen, so I look for the one I would most like to catch, picking out a beautiful third-year fish who is hanging under a clump of weed, showing himself from time to time as he moves to the left to feed. The first nymph up for trial – picked, I have to confess, at random – is the damselfly. It is bigger than most others in my selection, green, with tiny glass eyes and about an inch and a half long, half of which is a waggly tail. This is a tasty morsel for any fish and surely cannot fail with one as eager to feed as this one. But fail it does. After five casts not even a look, or any indication that the fish even acknowledged my fly's existence. Time to change. This time I go to the other extreme with a tiny olive nymph no more than a quarter of an inch long, barely bulkier than the thin wire of the hook onto which it is tied. The results are similarly dismal but yet more annoying, as the fish continues to feed just before or just after my fly has trundled past. Failure seems to make the sun burn hotter, so I rest back on my haunches to watch the fish to divine some clues. I cannot see what he is eating, but I can see how he is eating, only picking food that is in a layer 3 inches above the riverbed, the precise level at which he has positioned his body in the current. He doesn't look up. He doesn't look down. However he is willing to go a foot to the left for his chosen nymph, just so long as it is in the layer. Clearly the depth of my fly is critical.

Back in position behind the meadowsweet, this time I tie on a Pheasant Tail Nymph, a drab but simple pattern, so called because it is made with a few turns of brown pheasant feather held in place with copper wire. Its very ordinariness belies its effectiveness, the single most successful fly ever invented, which catches trout from Australia to Alaska and every continent in between. Created by a humble Wiltshire river keeper in the 1950s, on a chalkstream no more than 25 miles from where I stand today, it is a generic nymph pattern that must trigger something deep in the psyche of trout wherever they happen to live. My first cast has something of a tracer-fire quality to it. It would be good to make it my best, as I would a dry fly, but there is the variable of depth to add to direction and length. As my nymph lands on the surface I have the same problem as the real nymphs do in coming up; a taut layer of surface film that is hard to break through. Woefully inadequate as a first cast, the nymph has barely broken the surface by the time it passes, unnoticed, over the head of the trout. The second cast is a bit better; the nymph has taken on a little water and the extra density helps it sink maybe 4 or 5 inches. Not enough – I need to triple or quadruple that – but a start. With each successive cast I land the fly further and further ahead of the trout to give it time to sink. Depending on the refraction of the light or how it hangs in the water I can sometimes spot the nymph, but generally not, relying more on instinct (Zen?) than sight to estimate the depth as it passes the trout.

Soon the world narrows to just the cast, the nymph and the fish as I repeatedly lay out the line. Sometimes I think he has moved across to my fly. I tense, pull in a little extra line ready to strike by raising the rod tip the moment he opens and closes his mouth. But he doesn't take the fly, so I relax to let the line pass him by before re-casting. Other times I strike only to realize he had swallowed down a real nymph, not my apparently feeble imitation. As a rhythm establishes itself I have this absolute certainty that, just so long as I don't spook him with a splashy cast or careless shadow, I will catch this trout. It is simply a matter of timing: his and mine. And then it comes together. The cast. The movement of the line and the nymph. The posture of the trout as he readies himself. He moves to the left. He opens his mouth. I raise the rod tip. The rod bends, the line tightens and for a fraction of a moment I can see the confusion in the trout as the hook sets. Then anger replaces confusion as he violently shakes his head but unable to free himself heads away upstream, pulling line off the reel. Our trout is still young, without the wiles of the older trout who would have headed for a tree root or some other obstruction that spelt safety, but with his three-quarter-pound bulk he is easy to control, and within thirty seconds I guide him back downstream, into my hand; once released he heads back to the clump of weed where I first spied him, this time burrowing his head into it so that he disappears from view. Extraordinarily, none of the other fish seem fazed by what has occurred, continuing to feed and hold station as before. That is typically trout – do a bad cast or spook them in some way and the whole lot flee for cover. But hook one, have the ensuing fight and release it, and none of the others take a blind bit of notice.

On that basis I am tempted to try for another, but the day seems to be getting oppressive, hotter by the minute. Across the stubble of the meadows I can spy the Drowners House, and the siren call of its cool, damp interior beckons.

14
CAMS POINT

IT IS THE kingfisher, normally such a silent bird, that is my soundtrack as the autumn swallows up the summer. With every shortening day, from dawn to dusk, he whistles away in each successive corner of Gavelwood. To say he has a preset route would be wrong, but he does have a series of perches that he flits between. The fence posts at the cattle drink. The dead branches that hang out from the alders over Pike Pool. The handrail across Bailey Bridge. The seat in front of Drowners House. Even the 'Fishing Ends' sign at the boundary is a regular haunt. There are probably a dozen more he uses that I have not noticed as he criss-crosses between the meadows and streams. Quite why he has to make so much noise I have no idea – that flash of blue is more than enough to announce his presence – but I guess it is a territory thing. His call is more like a whistle, the sort you'd make by blowing short and sharp bursts through the gap of your middle teeth. Confusingly it also sounds a lot like the squeak with which the otters talk to each other, and sometimes at dusk or dawn I look around to see an otter only to catch the kingfisher blue out of the corner of my eye. The volume and frequency of his calls is quite at odds with his spring and summer persona when he barely muttered a word; I assume the nesting and nurturing took all his time back then. But now, freed of his family obligations, marking out his patch is the big priority. Winter is hard on kingfishers, so seeing off all possible competition before the going gets tough makes a certain amount of sense. It is not so much Darwinian survival of the fittest but survival of those who make the pre-emptive strike. The swans are a prime case in point.

There is by my reckoning plenty of space, weed and water for the parents and their three offspring to happily coexist all winter, but the bond with the six-month-old cygnets, still brown-feathered but close in size to the parents, breaks down overnight. One September evening they will be roosting together on the bank in a united family group, the next the cob is driving away his offspring. It is no gentle nudge to send them on their way; it is venomous hissing, biting and wing-beating aggression. The first two will take the hint and leave quickly; the runt of the litter hangs around for a few days longer, making increasingly forlorn attempts to ingratiate himself with his mother, until the attacks by his father make leaving a better option than staying. What fascinates me is how inestimably proud the cob seems to be of his achievement; for the next few days he will pump up and down the Evitt, chest out as he pushes a bow wave of water ahead of him, whilst the pen trails disconsolately behind, maybe regretting how little she fought for her cygnets back in the summer. One can only guess at the tone of the conversation should they ever talk.

Plenty of others are changing their lives as well; the swallows leave, the water voles secure their winter burrows and the bats seek out somewhere to hibernate, appearing less and less frequently with each passing evening. In the air, on the banks and in the fields the creatures and birds are laying in stores and hunkering down. Their best times are past until the spring, with winter survival now the only aim. But for the salmon and trout it is an altogether different story, their most important days still ahead. Under the gathering gloom of autumn the trout change in habit and appearance before my eyes, but the wait for the first salmon grows more anxious with each passing day.

Way down the Evitt, miles downstream of Gavelwood, past Middle Mill and in the brackish, tidal reaches where the pure chalkstream water and the sea mix, stands a white stone, arched bridge at Cams Point. It is very old. If you crane your head over the parapet the keystone is etched with the numerals seventeen hundred and something, the decade and year number damaged long ago so that they are indecipherable today. In my lifetime this was once a road bridge, the major route along the coast, but long since replaced by a concrete flyover and dual carriageway. Today the old bridge is a footpath and cycleway, beneath which the Evitt rushes out between the pillars on the ebb tide, in on the flood tide, with that strange hour betwixt high and low water when the surface looks oily, hanging there, going neither in nor out on the slack tide. Sometimes I break my journey to stop on the old bridge. It fascinates me that all life that migrates between the sea and the river must pass under this arch, no more than 20 yards across. Every eel, every sea trout and every salmon that tastes the chalkstream must pass through. I try to imagine how huge some of the fish that slipped in and out, unobserved by man, must have been. Surely some record-breakers? Plenty of fish will have made the journey in and out the once, but how many have crossed the Atlantic twice, three times, four times or more? The river hides more secrets than it reveals.

I cannot ever recall a time when I have been alone on the bridge; whether it be dawn, dusk or the pitch-black of night, there will always be someone fishing. It is not the fly-fishing of Gavelwood or even some practical spinning; this is hardcore bait fishing for food. The fast-moving water requires a particular technique not seen in many refined fishing books. Take a short, stout rod with a heavy-duty reel wound with thick fishing line. Attach to the end of the line a monstrous weight, with an extra piece of line with hook and bait tied at right angles about a yard above the weight. Poke your rod over the parapet, release the drag and let the weight pull the line off the reel, plopping into the water and then on for a few seconds until it rests on the bottom. Crank the reel to take up the slack, then wait whilst the hook and bait tugs on the tide a few feet above the riverbed ready to tempt an incoming or outgoing fish according to the flow.

The baits used are as varied as the people; you get the kids who while away the hours between fish (or no fish for that matter) listening to music or playing on their smartphones, whilst the old-timers indulge in a slow sort of conversational tennis that seems to cover every subject except fishing. They all have their baits of choice: huge balls of earthworms, prawns, luncheon meat, dead fish, you name it, at some point someone has probably tried it, but for all the variety this is far from being a fish every cast. In fact it is far from being a fish every day. Whenever I stop by and enquire, rarely does anyone seem to have caught anything. Actually more accurately nobody admits to catching anything. The best salmon run happens out of season, and as for a national fishing licence, well, best not to ask. Rookie Environment Agency bailiffs have been known to visit in the past, but they only do it the once. Eels are the most commonly caught fish, as well as the most disliked and commonly derided. Plenty of the bridge anglers will not even touch *Anguilla*, but there is an old guy who smokes them down on his allotment, so he steps forward to grab them, stuffing them into a hessian sack that gets more mobile as the day progresses. Sea bass are highly prized, sea trout less so, but really it is salmon everyone is after. I don't need to watch my river to know when the autumn salmon run has started; the number of anglers on the bridge tells me everything I need to know, and once I see them shoulder to shoulder in the early morning mist I know the arrival at Gavelwood is not so far away, spurring me on to put the river to bed before winter closes in.

It is a strange expression, but putting the river to bed is one of those phrases every river keeper uses, and as is the nature of the work we do, each of us has a slightly different take on the phrase. There is a sort of comforting finality to it; the jobs and tasks you have to complete to close the river once the fishing season ends. Some are relatively trivial housekeeping tasks like closing up the Drowners House, other critical to the well-being of the river to take it through winter and safely out to the spring the other side. There are all sorts of things to do; we make a final weed cut, trimming the weed to protect the weed from itself and the river as well. Big rafts of weed are vulnerable to heavy floods, which will tear it out by its roots from the riverbed. If that happens it will take years to re-establish, so we trim the weed back, a sort of pruning if you like, so the floods will flow harmlessly over and around it. The banks of weed we so carefully left in place in the height of summer to hold back the flow are now redundant, ready to be cut away to allow free passage for the water. In some places we will cut the weed in such a way as to redirect the flow to wash away an unwanted bed of silt that has built up or to prevent a reed bed encroaching any further out into the river. The boards in the hatches are removed and the sluice gates opened wide. It is all about letting the river pass through at speed, scouring out the unwanted accumulations of summer and refreshing the gravel riverbeds in time for spawning and the new life that will be created along the riverbed.

The last grass cut done, the scythes and mowers put away, we gather as a team for the most laborious of jobs, gravel-blasting. Years ago, long before my time, there was a certain romance to this, when the work was done by a plough horse and harrow, dragging the tines through the gravel bed to break up the hardening surface. Today we use high-powered water pumps to create soft plumped pillows of gravel at strategic spawning places around Gavelwood, but it is tedious work to the extreme. Fire up the pump, get into the river, push the steel probe into the riverbed to a depth of a foot, and hold it there whilst the water jet separates, dislodges and washes away the silt that has cemented the gravel stones together. When the water runs clean remove the probe and push it back into the next hard section. And so it goes on. For days. There was a time when we could rely on groups of students studying fish science at the local agricultural college turning up for a week of 'work experience', but even they seem to have wised up to the monotony. It is one of those tasks easy to talk yourself out of in the short term, but years down the line you'll reap the whirlwind as the wild fish population spins into decline.

Though we do the blasting mostly with spawning in mind, the gravel on the riverbed deserves more attention than we usually give it, being one of those building blocks around which the chalkstream world revolves. As I plunge the probe down for the umpteenth time all manner of life shows itself. The nymphs are ever-present; big mayfly nymphs are the most visible, darting around oblivious to the changing season, or the occasional bloodworm will twist about, discomfited by the commotion, until he comes to rest again on the riverbed. The jetting displaces shrimps in their hundreds; the trout and grayling downstream of where we are working are in for a heyday. Little snails and mussels tumble a few feet before finding a new home. Best of all, the truly ugly bullheads, aggressive at the best of times, definitely take umbrage at having their patch disturbed. For if anyone is king of the gravel bed it is the miller's thumb.

It scuds out from the stones under which it hides, to bite everything that passes. If it is food it is eaten. If it is another male it is escorted, nightclub bouncer style, away. If it is a female during the spring breeding season she is pulled into the gravel nest where the eggs will be laid. In common with that other aggressive guy the stickleback, the male bullhead guards the eggs until they hatch, but the offspring leave the nest swiftly before they can be eaten by their father. Charming, but it gets worse. Many river keepers regard them as vermin; voracious eaters of trout and salmon eggs, which by virtue of where they live seems to be a charge that sticks. In short there is nothing beautiful about the bullhead, in demeanour or even appearance, with its brown, mottled camouflage skin and oversized head. Can there be anything good to be said about a fish that is so prevalent yet little observed? With its ability to hide and bottom-dwelling habits nobody really sees it, but in weight terms it is reckoned to account for one quarter of all the fish in the Evitt, or any chalkstream for that matter. When you consider each bullhead only weighs a fraction of an ounce, that is an awful lot of fish. But as ever in a river, the one thing all fish love to eat is other fish, with the trout getting their revenge, hoovering up the bullheads, fat from the protein-rich eggs, to provide the trout with four-fifths of their diet over the winter. Nature often has a great way of evening things up.

By late September the sense that the best is behind us and the worst ahead seems to pervade Gavelwood. With each passing day I get the urge to fish more, in case it might be my last. Way back in the summer I passed up too many opportunities, knowing that there were many more days to come. This time it is different. Today was the first time in months that I felt a tangible shiver of cold, enough to warrant a jacket and to discourage dawdling on my morning rounds to clear the sluices and check the hatches. The Drowners House felt warm inside instead of cool as I collected my rod for the last time. The windows were obscured by condensation, the rafters above strewn with cobwebs as even the spiders got in on the last-minute feeding act. The record book showed nobody had been fishing all week; things must be slowing down. Back outside I was suddenly struck by the brown tinge of the trees; gone is the vivid green of summer, replaced with leaves that have started to crinkle and darken around the edges, with drifted piles on the ground.

But most of all it is the trout that have changed – by now they have seen it all, but driven by a survival imperative they are ready to feed when the right thing comes along. They don't skulk in the deep pools. They don't dash away at the slightest sign of a person. They don't steadfastly refuse all offerings. These are fish that know they have to feed heavily before the winter shuts down the food chain. Fish do definitely act differently as the days start to shorten; they are much bolder in the water, sitting higher, more obviously visible and less easily spooked. Eager though they are to eat as much as they can, they do get choosy. On days like this the fish will follow my fly for feet or even yards, swimming directly under it, at some points almost touching it with his nose. It is not so much fishing, more a game of high-stakes poker as the fish eyes my fly, willing it to be real, but sensing it may not be. This is not the smash-and-grab of the mayfly hatch or the delicate sipping of a tiny fly from the surface on a blazing hot day, but rather a war of attrition. I cast, the fish follows. I cast again, this time twitching the fly to imitate a clumsy sedge, and the fish comes closer still. I try a cast just on the periphery of the trout's vision, and sure enough, his curiosity aroused, he tracks across to take a look, but at the last moment declines to eat. I should be exasperated, but the more he follows, the more the adrenalin pumps through my veins. I will him to open that mouth, suck down the fly and let me raise my rod to strike in victory. But time and again he calls my bluff. I change the fly. I put on a new tippet. I move a few steps along the bank to alter the drift of my fly. We are becoming the Cold War warriors who can't bear to give up, but neither is willing to make the first strike. Each time the fish turns away from the fly I am compelled to cast again, and for a while he follows every cast, until he bores of my offering. Clearly some sort of détente is called for, so taking the hint I retreat to a seat on the bank as the watery sun brings a glimmer of brightness to the morning.

After a while it gets really quite warm, so I shed my jacket. The moorhen chirrups. The tops of the reeds in the river margin move as if by some unseen hand, which means the water voles are gathering winter supplies for sure. The swans push by without stopping on their daily patrol to warn and ward off interlopers. The cattle are tearing at the last of the grasses, the meadows close to being grazed out. A flight of ducks land for a moment, think the better of it and quacking with some sort of righteous indignation that I don't understand, take off as quickly as they arrived. A cloud of tiny olives magically appear without warning or indication of where they came from. They momentarily envelop me as they move along the bank, so I swat them gently away as you would smoke from a campfire. The river flows on by as it has all year.

Soon my trout starts to feed again, establishing a slow rhythm as he rises porpoise-like revealing just his back and dorsal fin, taking hatching nymphs that are caught in the surface film. A last cast for a last fish? It hardly seems to be right when we have both come so far, so I snip off my fly, reel in my line and head off.

EPILOGUE

A FEW DAYS later come the first heavy rains of the winter, swelling the river and pushing the scent of the Evitt through the estuary into the sea, those few tiny parts in a billion washing past the salmon waiting out in the English Channel. Triggered by the smell, that deep-seated desire to return home urges them forward. After many months and many thousands of miles of travel this siren call impels our *salar* on the final leg of his journey. Along the coast he comes, into the estuary, under the bridge at Cam's Point and on to the pool beneath Middle Mill, where he pauses. But the early arrivals don't pause for long. The rapid change from salt to fresh water seems to cause them no difficulty. Our fish, his strong silver body fresh from the sea, effortlessly pushes up through the fast torrent of the Middle Mill hatch to continue upstream. Night and day he swims, eating up the miles until at Pike Pool, unseen, he halts at the very place of his birth, six years and four thousand miles ago.

As I head along the river the darkness of the late November afternoon closes in, a gust of wind blowing a litter of dead leaves off the alders and onto Pike Pool. A group of ducks fly fast across the sky, seeking sanctuary before night falls. The meadows are strangely silent, the cattle absent now the fields are getting wet underfoot. The bats no longer flit in the dusk, but two owls fill the silence hooting to each other across the length of Gavelwood as the mist rolls in. Ahead of me *salar* rises up from the river through the mist, leaping from the water, his bright body catching the rays of the early rising moon, a crack resonating across the valley as he crashes back into the water. He jumps just the once. Maybe it is the exuberance of being home. Or maybe he is trying to signal his presence to me. Whichever it may be, it is enough to send me home happy, content that the river year has come full circle.

BIBLIOGRAPHY

Glasspool. Jim, *Chalk Streams* (Romsey, Test & Itchen Association, 2007)

Halford, Frederic, *Dry-Fly Fishing in Theory & Practice* (London, John Bale & Sons, 1888)

Herd, Andrew, *The History of Fly Fishing* (Ellesmere, Medlar, 2011)

Kite, Oliver, *Nymph Fishing in Practice* (Shrewsbury, Swan Hill Press, 1977)

Lapsley & Bennett, *Matching the Hatch* (Ludlow, Merlin Unwin, 2010)

Maxwell, Sir Herbert, *British Fresh-Water Fishes* (Camberley, Hutchinson & Co, 1904)

Pease, R H, *The River Keeper* (Newton Abbot, David & Charles, 1982)

Rangeley-Wilson, Charles, *Chalkstream* (Ellesmere, Medlar, 2009)

Sawyer, Frank, *Keeper of the Stream* (The Country Book Club, 1954)

Seymour, Richard, *Fishery Management and Keepering* (London, Charles Knight & Co, 1970)

Vines, Sidney, *The English Chalkstreams* (London, Batsford, 1992)

Wilson, Dermot, *Fishing the Dry Fly* (London, Unwin Hyman, 1987)

ACKNOWLEDGEMENTS

This is true. One morning nearly three years ago two emails popped into my inbox from two entirely unconnected people, both of which essentially said the same thing: isn't it high time you wrote a book about the chalkstreams. Sometimes you cannot escape your fate.

One of those two emails came from Emma Kirby, an old friend, who was re-starting her literary agency after time out to raise a family. With the germ of an idea we started to hone a synopsis but soon got into fights about my 'style'. A referee was required, so we tied up with freelance editor Imogen Fortes to adjudicate. Without having been caught in the pincer of these two talents I can say with complete certainty *Life of a Chalkstream* would never have happened. To you both I owe a huge debt.

Make no mistake, the image of the tortured writer in a lonely attic scribbling away is a very long way from the truth, at least for me. Books are collaborative and I was fortunate to wash up on the shores of HarperCollins with Myles Archibald, the publisher, and his incredibly supportive team. From the original manuscript they have honed, polished and delivered a book of which I hope they are proud – they should be – it's a great job.

Many others have contributed to this book. Shaun Leonard, Director of the Wild Trout Trust put his head on the block and kindly read the manuscript to point out my errors. Any that remain are mine and solely mine, as are the opinions and practices that are not entirely in tune with those of the Trust. Denise Ashton, also of the Trust, let me raid her contacts book. Andy Heath at the Derbyshire Rivers Trust provided the view from the north. At the Environment Agency Heb Leman has always extended my education on all things river, and Lawrence Talks probably knows more about chalkstreams than any other living being. Martin De Retuerto at the Hampshire Wildlife Trust provided the living history. My knowledge of trout rearing is in large part due to Trevor Whyatt at Allenbrook in Dorset.

My livelihood for the past twenty years plus has been made by conserving chalkstreams and organising fly-fishing trips to these very special rivers. The day-to-day logistics of hosting thousands of anglers each year, one hundred and twenty miles of river plus the fishing school here at Nether Wallop Mill in Hampshire from April to October is sometimes quite mind-boggling, but at Fishing Breaks I have the most amazing team who make it happen. Diane Bassett is the linchpin in the office, creating enough space in my days to let me actually do some writing. Jonny Walker and Kelly Hewlett make sure the rivers and cabins are perfect to the eye. My long-time fishing guides, some now retired, Duncan Weston, Simon Ward, Marcus McCorkell,

John Stephens, Alan Middleton, Tony King and Mark Bedford-Russell have all taught me stuff I did not know and still make every new day on the river a pleasure.

Each day of the year I meet or talk with the river keepers, owners and guides on the dozens of chalk-streams from Dorset in the west to Yorkshire in the north with whom I have connections, all of whom have always shown me great trust and shared their knowledge unstintingly. To name names would be invidious to those left unnamed and to compound the crime I would make the cardinal error of missing someone out. So to you all, including those who have gone to the great river in the sky, take heart that there is a bit of each and every one of you, plus your river, in this book.

Closer to home I owe eternal thanks to Karrie for the support and the chance to jack in a perfectly good career all those years ago to start what I have today. And finally to Minnie my daughter and Mary the biggest thank-you of all; this would be pointless without you.

Nether Wallop Mill, Hampshire, April 2014

INDEX